Statforms

'Pro-formas' for the guidance of statistical calculations

H. C. Dawkins

Department of Forestry
University of Oxford

Dawkins - Statforms

ERRATA

Statform 12A
Line 33, total should read totals.

Statform 12B
Line 36/37, negative sign preceding ac should be positive.

Statform 23
Line 27, column B, the figure 10 in the worked example should be 7.

Edward Arnold

© H. C. Dawkins 1975

First published 1975
by Edward Arnold (Publishers) Ltd.,
25 Hill Street
London W1X 8LL

ISBN: 0 7131 2359 X

Printed in Great Britain by
Whitstable Litho., Whitstable, Kent

Introduction

These statforms are intended to provide a series of 'pro-formas' for the guidance of statistical calculations on small sets of observations, especially those that arise in biological sampling and experiment. They do not form a text-book, nor a course, nor do they give any guidance as to which technique is suitable for any particular problem. Such guidance can be found in texts such as R. E. Parker's *Introductory Statistics for Biology* (Edward Arnold, 1973), or the two works cited below.

The statforms method has evolved over a period of twenty years, firstly in silvicultural research and more recently during statistical courses, tutorial work and graduate supervision in the Oxford University Departments of Forestry, Botany, Agriculture and Zoology. They first appeared in print in Paper No. 41 of the Commonwealth Forestry Institute, 1968. At this stage the context was solely that of forest and agricultural research, with most tests and confidence limits based on the analysis of variance. The series lacked non-parametric methods and any direct application of binomial or Poisson distributions. Over the past seven years some of these have been included, in response to the needs of users outside the earthy biological sciences, especially, for instance, those dealing with behavioural, psychological and physiological observations. As well as the additions, the first-series forms have been re-drawn and their texts re-written in the light of class and individual experience.

The statforms have two main uses :

(1) in statistical education, allowing the teacher to keep track of progress and arithmetical performance, by reference to the numbered lines and headed columns on the forms. The student is then left with work in a form clear enough for later reference and revision.

(2) by individual workers, including students, who may be rusty in technique but sufficiently educated to know that a certain statistic or test is appropriate. Again his work is forced into a form easy to check and simple to refer to later.

Each statform is printed in two modes : as a blank and as a filled-in worked example. The blank is intended for use, the worked example provides guidance. Most worked examples are taken from the 6th edition of Snedecor and Cochran's very widely available *Statistical Methods* (Iowa State University Press, 1967 et seq.) which provides zoological, botanical and general physiological problems, from the 2nd edition of Campbell's *Statistics for Biologists* (Cambridge University Press, 1974), and some from Parker's book already cited. I have to acknowledge with thanks the permission of all three publishers for the use of their material, and also the permission of Oxford University for this revision of the Institute Paper. My thanks are also due to J. N. R. Jeffers, J. F. Scott and F. H. C. Marriott for comments and help with the embryonic stages of some of the statforms, in one case from as early as 1958.

To avoid what would otherwise have been a very tiresome job, text in the present series of statforms has been handwritten rather than typed, by Mr Bill Rides of the Oxford Department of Forestry. The numerical examples were entered by Miss Linda Roberts of the same Department. I am very grateful to both for their painstaking work on the series, and on several other statforms that for reasons of space have not been included here.

H. C. Dawkins
Oxford, 1975

Contents

STATFORM 1

#		
1	STATISTICS OF THE NORMAL DISTRIBUTION *and*	
2	SAMPLING ERROR FROM SMALL RANDOM SAMPLES	
3	with provision for two sets of observations	
4	Nature of the observations :	
5	A :	
6	B :	

Listing of observations

	A	B
7.		
8		
9		
10		
11		
12		
13		
14	Sum of the observations : Σy =	
15	Number of observations : n =	
16	Mean : $\Sigma y/n$, $\bar{y}$ =	
17	Squared sum : $(\Sigma y)^2$ =	
18	Sum of squared observations : Σy^2 =	
19	less correction for mean : $(\Sigma y)^2/n$ =	
20	gives sum-of-squares : SSy =	
21	Degrees of freedom : $n-1$, $d.f.$ =	
22	Variance of the observations : $SSy/d.f.$, Vy =	
23	Standard deviation : $\sqrt{Vy}$, SDy =	
24	Coefficient of variation % : $100 \times SDy/\bar{y}$, $CV\%$ =	
25	Sampling % expressed as a proportion : f =	
26	Unsampled residue of population : $1-f$, u =	
27	Variance of mean : $u \times Vy/n$, $V\bar{y}$ =	
28	Standard error of mean : $\sqrt{V\bar{y}}$, $SE\bar{y}$ =	
29	Student's t at P.05 (or other level) : t =	
30	Confidence range : $\pm t \times SE\bar{y}$, C =	
31	Confidence limits { Upper : $\bar{y} + C$ =	
32	around the mean { Lower : $\bar{y} - C$ =	
33	Sampling error % $100 \times C/\bar{y}$: $E\%$ =	
34	Estimation of sample size for a given	
35	sampling error. If required sampling error $e\%$ =	
36	then no. of sample } approx : $\left(\dfrac{CV\% \times t}{e\%}\right)^2$ =	
37	units required	

STATFORM 1

#		A	B
1	STATISTICS OF THE NORMAL DISTRIBUTION *and*		
2	SAMPLING ERROR FROM SMALL RANDOM SAMPLES		
3	with provision for two sets of observations		
4	Nature of the observations : *		
5	A : 17 samples of vitamin C concentration.		
6	B : Weight gains of 10 pigs .		

Listing of observations

A

16	22	17	16
22	17	18	23
21	15	29	
20	13	17	
23	22	22	

B

32	19	19
31	24	30
11	53	
30	44	

#			A	B
14	Sum of the observations :	Σy =	333	293
15	Number of observations :	n =	17	10
16	Mean : $\Sigma y/n$,	$\bar{y}$ =	19.6	29.3
17	Squared sum :	$(\Sigma y)^2$ =	110889	85849
18	Sum of squared observations :	Σy^2 =	6773	9949
19	less correction for mean :	$(\Sigma y)^2/n$ =	6522.88	8584.9
20	gives sum-of-squares :	SSy =	250.12	1364.1
21	Degrees of freedom : $n-1$,	d.f. =	16	9
22	Variance of the observations : $SSy/d.f.$,	Vy =	15.63	151.57
23	Standard deviation : $\sqrt{Vy}$,	SDy =	3.95	12.31
24	Coefficient of variation % : $100 \times SDy/\bar{y}$,	$CV\%$ =	20.2	42.0
25	Sampling % expressed as a proportion :	f =		0.1
26	Unsampled residue of population : $1-f$,	u =		0.9
27	Variance of mean : $u \times Vy/n$,	$V\bar{y}$ =		13.64
28	Standard error of mean : $\sqrt{V\bar{y}}$,	$SE\bar{y}$ =		3.69
29	Student's t at P.05 (or other level) :	t =		2.26
30	Confidence range : $\pm\, t \times SE\bar{y}$,	C =		8.34
31	Confidence limits {Upper : $\bar{y} + C$	=		37.64
32	around the mean {Lower : $\bar{y} - C$	=		20.96
33	Sampling error % $100 \times C/\bar{y}$:	$E\%$ =		28.5
34	Estimation of sample size for a given			
35	sampling error. If required sampling error $e\%$ =			10%
36	then no. of sample } approx : $\left(\dfrac{CV\% \times t}{e\%}\right)^2$ =			
37	units required			90

* Snedecor & Cochran, Statistical Methods, Iowa State University Press 1967, A, p.48-9 ; B, p.69.

STATFORM 2

1	STATISTICS OF THE NORMAL DISTRIBUTION
2	for large random samples which can conveniently
3	be grouped in a frequency table.
4	Nature of the observations :
5	

6 (1)Enter value of each size-class in column y, in order of magnitude
7 increasing downwards. (2)Enter observations as frequencies in column f.
8 (3) Select the median class (approx), enter zero in column d on median
9 class line. Also in column d, enter the deviates −1, −2, etc. upwards
10 from the zero and +1, +2 etc. downwards until all size-classes are
11 occupied. Complete remaining columns as headed.

	Size-class y	Frequency f	Deviations d	$f \times d$	d^2	$f \times d^2$
12						
13						
14						
15						
16						
17						
18						
19						
20						
21						
22						
23						
24						
25						
26						
27	$n = \Sigma f =$		$\Sigma (f \times d) =$		$\Sigma (f \times d^2) =$	

28 Class width or interval: $C = $ ....

29 Mid-point of selected median class : $y_0 =$

30 Calculation of standard deviation.   Calculation of mean :

31 $\dfrac{[\not{z} (f \times d)]^2}{} \qquad =$ $\not{z} (f \times d)/n \qquad = $ ....

32 $\not{z} (f \times d^2) = $ ....   Multiply by C to give: ....

33 $[\not{z} (f \times d)]^2/n \qquad = $ _____   Add y_0 } $\bar{y}$ =

34 by subtraction : $SSd = $ _____   to give)

35 $SSd/(n-1)$, $Vd = $ .... $SDd = \sqrt{Vd} = $

36 Standard deviation : $= C \times SDd \qquad = $ $= SDy.$

37 Continue as on Statform 1 line 23 onwards

STATFORM 2

1	STATISTICS OF THE NORMAL DISTRIBUTION
2	for large random samples which can conveniently
3	be grouped in a frequency table.
4	Nature of the observations :
5	**Measurements on a sample of 30 individuals.** *

6 (1)Enter value of each size-class in column y, in order of magnitude
7 increasing downwards. (2)Enter observations as frequencies in column f.
8 (3) Select the median class (approx), enter zero in column d on median
9 class line. Also in column d, enter the deviates −1, −2, etc. upwards
10 from the zero and +1, +2 etc. downwards until all size−classes are
11 occupied. Complete remaining columns as headed.

	Size-class y	Frequency f	Deviations d	$f \times d$	d^2	$f \times d^2$
14	10	1	−4	−4	16	16
15	11	1	−3	−3	9	9
16	12	3	−2	−6	4	12
17	13	7	−1	−7	1	7
18	14	8	0	0	0	0
19	15	6	1	6	1	6
20	16	3	2	6	4	12
21	17	1	3	3	9	9
22						
23						
24						
25						
26						
27	$n = \Sigma f = $ 30		$\Sigma (f \times d) = $ −5		$\Sigma (f \times d^2) = $ 71	

28 Class width or interval: $C = $ 1·0
29 Mid-point of selected median class: $Y_0 = $ 14

30 Calculation of standard deviation.   Calculation of mean:
31 $[\Sigma (f \times d)]^2$ = 25 $\Sigma (f \times d)/n$ = −0·17
32 $\Sigma (f \times d^2) = $ 71   Multiply by C to give: −0·17
33 $[\Sigma (f \times d)]^2/n$ = ·83   Add Y_0 } $\bar{y}$ = 13·83
34 by subtraction : $SSd = $ 70·17   to give)
35 $SSd/(n-1)$, $Vd = $ 2·42 $SDd = \sqrt{Vd} = $ 1·56
36 Standard deviation : $ = C \times SDd$ = 1·56 $ = SDy.$

37 Continue as on Statform 1 line 23 onwards

* R. E. Parker , Introductory Statistics for Biology , Edward Arnold 1973 , p. 90 .

1	STRATIFIED RANDOM 2-PER-BLOCK SAMPLING
2	*Blocks of uniform size sampled by equal-sized pairs of sample units.*
3	*Nature of the observations :*
4	*Blocks :*
5	*Sample units :*

6	No. of blocks : $B =$	No. of sample units : $N = 2B =$

7	Sampling proportion: (ie. sample % ÷ 100) = ...
8	∴ Unsampled residue of population : $U =$

Lesser variate method

9 / 10	Block No.	Observations Y_1 Y_2	Difference $Y_1 - Y_2$	Difference ^{2}	Lesser Y_1 or Y_2	Probability
11						·25
12						·17
13						·12
14						·09
15						·06
16						·05
17						·04
18						·03
19						·02
20						·02
21						·02
22						·02
23	$\Sigma y =$	$\Sigma (Y_1 - Y_2)^2 =$	....		= Sum
24	Mean, $\bar{y} =$			....	= Mean

25	Divide by N^2 (=) to give $V\bar{y}$	= ....
26	Multiply by U: $V\bar{y} \times U$	= ....
27	Square-rooted to give : $SE\bar{y}$	= ....
28	Student's t at d.f. = B for $\}$ t	=
29	a probability of $P =$	....
30	Confidence range, $\pm t \times SE\bar{y}$, C	= ....
31	Confidence limits : $\{$ Upper, $\bar{y} + C$	= ....
32	$\{$ Lower, $\bar{y} - C$	= ....
33	Sampling error percent = $100 \times C/\bar{y}$	= .... = E%

Mean of the lesser variates is a lower confidence limit with a probability fixed by the number of blocks sampled

34	Estimation of sample size for a given sampling error :
35	e.g. if required sampling error is $e\%$ = ...
36	No. of sample units $\}$ approx. = $\left(\dfrac{E\%}{e\%}\right)^2 \times N =$
37	required $\}$ ....

STATFORM 3

STRATIFIED RANDOM 2-PER-BLOCK SAMPLING

#	
1	STRATIFIED RANDOM 2-PER-BLOCK SAMPLING
2	Blocks of uniform size sampled by equal-sized pairs of sample units.
3	Nature of the observations : Sampling two per herd from
4	Blocks : Nine herds of milking cows } total of
5	Sample units : Individual cow milk yields } 270 cows. *
6	No. of blocks : B = 9	No. of sample units : N = 2B = 18

Line	Block No.	Observations y_1	Observations y_2	Difference $y_1 - y_2$	Difference^2	Lesser variate method — Lesser y_1 or y_2	Probability
7	Sampling proportion: (ie. sample % ÷ 100) = 0.067						
8	∴ Unsampled residue of population : U = 0.933						
11	1	780	816	−36	1296	780	·25
12	2	803	834	−31	961	803	·17
13	3	776	818	−42	1764	776	·12
14	4	875	929	−54	2916	875	·09
15	5	857	828	29	841	828	·06
16	6	774	813	−39	1521	774	·05
17	7	841	730	111	12321	730	·04
18	8	774	781	−7	49	774	·03
19	9	742	790	−48	2304	742	·02
20							·02
21							·02
22							·02

Line		
23	Σy = 14561 $\qquad \Sigma (y_1 - y_2)^2$ = 23973	7082 = Sum
24	Mean, $\bar{y}$ = 808·9	787 = Mean
25	Divide by N^2 (= 81) to give $V\bar{y}$ = 295·96	Mean of the lesser
26	Multiply by U: $V\bar{y} \times U$ = 276·13	variates is a lower
27	Square-rooted to give : $SE\bar{y}$ = 16·62	confidence limit with
28	Student's t at d.f. = B for } t =	a probability fixed
29	a probability of P = ·05 } 2·26	by the number of
30	Confidence range, $\pm t \times SE\bar{y}$, C = 37·56	blocks sampled
31	Confidence limits : { Upper, $\bar{y} + C$ = 846·5	
32	{ Lower, $\bar{y} - C$ = 771·3	
33	Sampling error percent = $100 \times C/\bar{y}$ = 4·6 = E%	
34	Estimation of sample size for a given sampling error :	
35	e.g. if required sampling error is e % = 3%	
36	No. of sample units } approx. = $\left(\dfrac{E\%}{e\%}\right)^2 \times N$ =	
37	required 42	

* R.C.Campbell , Statistics for Biologists , C.U.P 1974
p.374 , Area 1 , Breed B.

STATFORM 4

			Totals
1	**STRATIFIED RANDOM SAMPLING**		
2	Sample units of equal size in blocks of differing sizes.		
3	Nature of the observations :		
4	Blocks :		
5	Sample units :		
6	Block number : ___ ___ ___ ___ ___ ___ ___		Totals
7			
8	Observations of		
9	y listed under		
10	each block column		
11			
12			
13	Totals : $\Sigma y =$		
14	No. of units : $n =$		
15	Block means : $\bar{y} =$		ΣW
16	Block size : $W =$		
17	Sampling proportion, $f =$		
18	Unsampled proportion, $u =$		$\Sigma (d.f.)$
19	$n-1,$ $d.f. =$		
20	$\Sigma y^2 =$		
21	$(\Sigma y)^2/n,$ $CF =$		
22	$\Sigma y^2 - CF,$ $SSy =$		
23	$SSy/d.f.,$ $Vy =$		
24	$Vy/n,$ $V'\bar{y} =$		$\Sigma (W^2 V\bar{y})$
25	$V'\bar{y} \times u,$ $V\bar{y} =$		
26	$V\bar{y} \times W^2,$ $W^2 V\bar{y} =$		
27	$W \times \bar{y},$ $W\bar{y} =$		
28	Summing over all blocks :		$\Sigma (W\bar{y})$
29	Weighted mean $= \dfrac{\Sigma (W\bar{y})}{\Sigma W} =$ _____ $=$ .... $\bar{Y}$		
30			
31	Variance of weighted mean $= \dfrac{\Sigma (W^2 V\bar{y})}{(\Sigma W)^2} =$ _____ $=$ .... $V\bar{y}$		
32			
33	Standard error of weighted mean $= \sqrt{V\bar{y}} =$ $=$ $SE\bar{y}$		
34	Confidence range and sampling error as on Statform 3 lines 30–33,		
35	but d.f. will be $\Sigma (d.f.)$ from line 19 above only if block variances		
36	(line 23) are homogenous. If not so, overall d.f. will be less than		
37	$\Sigma (d.f.)$ but more than the number of blocks.		

STATFORM 4

			1A	1B	1C	2A	2B	2C		Totals
1	STRATIFIED RANDOM SAMPLING									
2	Sample units of equal size in blocks of differing sizes.									
3	Nature of the observations: Milk yield 100 gall. units. *									
4	Blocks: Six blocks as shown, pages 374-6.									
5	Sample units: Individual yields, first columns only.									
6	Block number:		1A	1B	1C	2A	2B	2C		Totals
7	Observations of y listed under each block column		9.5	7.5	6.9	9.5	7.5	5.5		
8			9.8	8.0	6.3	9.0	6.9	6.7		
9			9.7	8.1	7.7		6.4	6.3		
10				8.6						
11										
12										
13	Totals: Σy =		29.0	32.2	20.9	18.5	20.8	18.5		
14	No. of units: n =		3	4	3	2	3	3		
15	Block means: $\bar{y}$ =		9.67	8.05	6.97	9.25	6.93	6.17		ΣW
16	Block size: W =		2.0	3.0	2.5	2.0	2.0	2.0		13.5
17	Sampling proportion, f =		0.15	0.13	0.12	0.10	0.15	0.15		
18	Unsampled proportion, u =		0.85	0.87	0.88	0.90	0.85	0.85		$\Sigma(d.f.)$
19	$n-1$, $d.f.$ =		2	3	2	1	2	2		12
20	Σy^2 =		280.38	259.82	146.59	171.25	144.82	114.83		
21	$(\Sigma y)^2/n$, CF =		280.3	259.21	145.60	171.125	144.213	114.083		
22	$\Sigma y^2 - CF$, SSy =		.0467	0.61	0.987	0.125	0.607	0.747		
23	$SSy/d.f.$, Vy =		.0233	.2033	.493	0.125	0.303	0.373		
24	Vy/n, $V'\bar{y}$ =		.0078	.0508	.164	.0625	0.101	.124		
25	$V'\bar{y} \times u$, $V\bar{y}$ =		.0066	.042	.1447	.0562	.086	.106		$\Sigma(W^2 V\bar{y})$
26	$V\bar{y} \times W^2$, $W^2 V\bar{y}$ =		.0264	.398	.904	.225	.344	.424		2.321
27	$W \times \bar{y}$, $W\bar{y}$ =		19.34	24.15	17.43	18.5	13.86	12.34		105.62
28	Summing over all blocks:									$\Sigma(W\bar{y})$

29-30 Weighted mean = $\dfrac{\Sigma(W\bar{y})}{\Sigma W}$ = $\dfrac{105.62}{13.5}$ = ...7.82....$\bar{Y}$

31-32 Variance of weighted mean = $\dfrac{\Sigma(W^2 V\bar{y})}{(\Sigma W)^2}$ = $\dfrac{2.321}{182.25}$ = 0.013 $V\bar{y}$

33 Standard error of weighted mean = $\sqrt{V\bar{y}}$ = 0.114 = $SE\bar{y}$

34 Confidence range and sampling error as on Statform 3 lines 30-33,
35 but d.f. will be $\Sigma(d.f.)$ from line 19 above only if block variances
36 (line 23) are homogenous. If not so, overall d.f. will be less than
37 $\Sigma(d.f.)$ but more than the number of blocks.

* R.C. Campbell, Introductory Statistics for Biology, C.U.P. 1974.

	# t-TEST OF THE DIFFERENCE BETWEEN THE MEANS OF TWO SETS OF PAIRED OBSERVATIONS.			
1				
2				
3		Observations		Differences
4	*Nature of the observations*	**A**	**B**	$D = A - B$
5				
6	A :			
7				
8				
9				
10	B :			
11				
12				
13				
14				
15				
16				
17	*Number of pairs :*	Sums :	ΣA	ΣB	ΣD
18				
19	$n =$	Means:	$\bar{A}$	$\bar{B}$	$\bar{D}$
20				

21	Sum of squared differences :	$\Sigma D^2 =$	....
22	Correction factor : $(\Sigma D)^2/n$ = .../.......	=	_____
23	Sum-of-squares, by subtraction,	SS =	
24	Degrees of freedom : n-1,	d.f. =	....
25	Variance of the differences : SS/d.f.,	V_D =	....
26	Variance of the mean difference, V_D/n,	$V_{\bar{D}}$ =	....
27	Standard error of the difference, $\sqrt{V_{\bar{D}}}$,	$SE_{\bar{D}}$ =	....
28	t-ratio of the difference, $\bar{D}/SE_{\bar{D}}$, t-ratio	=	
29	Student's t for above d.f. at e.g. P·05 is t	=	....

30 If the calculated t-ratio exceeds t then the difference $\bar{D}$ is "significant"
31 If the difference could only reasonably be in one direction and only one
32 is to be tested, compare t-ratio with the value of Student's t at
33 twice the above P, i.e. P ·10 ("one tailed test").

34	Confidence limits for $\bar{D}$ are given by $\pm t \times SE_{\bar{D}}$	=	.... $= C$
35			
36	Confidence range	Upper : $\bar{D} + C =$	....
37		Lower : $\bar{D} - C =$	....

STATFORM 5

t-TEST OF THE DIFFERENCE BETWEEN THE MEANS OF TWO SETS OF PAIRED OBSERVATIONS.

Nature of the observations	Observations A	Observations B	Differences $D = A - B$
A: **Preparation 1** (of virus extract)	31	18	13
	20	17	3
	18	14	4
	17	11	6
B: **Preparation 2**	9	10	-1
	8	7	1
observations of lesions	10	5	5
on the two halves of	7	6	1
eight tobacco leaves *			

Number of pairs : $n =$ **8**	Sums :	ΣA 120	ΣB 88	ΣD 32
	Means :	$\bar{A}$ 15	$\bar{B}$ 11	$\bar{D}$ 4

Sum of squared differences :	$\Sigma D^2 =$	258
Correction factor : $(\Sigma D)^2 / n$ = 1024/ 8	=	128
Sum-of-squares, by subtraction,	SS =	130
Degrees of freedom : $n-1$,	d.f. =	7
Variance of the differences : SS/d.f.,	$V_D =$	18·57
Variance of the mean difference, V_D/n,	$V_{\bar{D}} =$	2·32
Standard error of the difference, $\sqrt{V_{\bar{D}}}$,	$SE_{\bar{D}} =$	1·52
t-ratio of the difference, $\bar{D}/SE_{\bar{D}}$, t-ratio	=	2·63

Student's t for above d.f. at e.g. P ·05 is   t   =   **2·36**

If the calculated t-ratio exceeds t then the difference $\bar{D}$ is "significant"

If the difference could only reasonably be in one direction and only one is to be tested, compare t-ratio with the value of Student's t at twice the above P, i.e. P ·10 ("one tailed test").

Confidence limits for $\bar{D}$ are given by $\pm t \times SE_{\bar{D}}$	=	3·59 .... = C
Confidence range Upper : $\bar{D} + C$ =		7·6
Lower : $\bar{D} - C$ =		0·4

* Snedecor & Cochran, Statistical Methods, Iowa State University Press 1967, p. 95.

1	*t*-TEST OF THE DIFFERENCE BETWEEN THE MEANS
2	OF TWO SMALL INDEPENDENT SAMPLES A AND B
3	using their pooled mean variance
4	Nature of the observations
5	A :
6	B :

Listing of observations

A   B

		A	B
14	Sum of observations :	ΣA = ...	ΣB = ...
15	Number of observations :	n_A = ...	n_B = ...
16	Means : Σ/n :	$\bar{A}$ =	$\bar{B}$ =
17	Difference between means, $\bar{A} - \bar{B}$, D =		
18	Sum of squared observations	ΣA^2 = ...	ΣB^2 = ...
19	Correction factors :	$(\Sigma A)^2/n_A$ = _____	$(\Sigma B)^2/n_B$ = _____
20	Sums-of-squares :	SSA =	SSB =
21	Degrees of freedom : n-1 :	d.f.$_A$ = ...	d.f.$_B$ = ...
22	Variance : SS/d.f.	V_A =	V_B =

23	If variance ratio $\left\{\dfrac{\text{Greater } V}{\text{Lesser } V}\right.$ = ——— = ... $\left.\right\}$ exceeds the
25	F-table value at P ·025, the *t* test may be unreliable;
26	Use Mann-Whitney (Statform 22) or a transformation
27	Mean variance = $\dfrac{SS_A + SS_B}{d.f._A + d.f._B}$ = ——— = ... = Vm
29	Standard error $\left\{\right.$ = $\sqrt{Vm\left(\dfrac{1}{n_A} + \dfrac{1}{n_B}\right)}$ = ... = SE_D
30	of difference
31	t-ratio = $\dfrac{\text{Difference } D}{SE_D}$ = ——— = ... = t-ratio
33	Student's *t* for d.f. of (d.f.$_A$ + d.f.$_B$) at P .... is .... = t
34	If *t*-ratio exceeds Student's *t*, the difference D is "significant"
35	Confidence limits for D are given by $t \times SE_D$, = ... = C
36	Confidence range : $\left\{\begin{array}{l}\text{Upper} : D+C = ...\\ \text{Lower} : D-C =\end{array}\right.$

STATFORM 6

1	**t-TEST OF THE DIFFERENCE BETWEEN THE MEANS**
2	**OF TWO SMALL INDEPENDENT SAMPLES A AND B**
3	*using their pooled mean variance*

4	Nature of the observations	Tail length in mm. of
5	A : island A }	adult males of small
6	B : island B }	mammal. *

	Listing of observations	
7		
8	A	B
9	88 , 87, 86, 85, 86, 87	87 , 85, 85, 86 , 84, 85
10	86, 87, 85, 86, 86	84, 86, 83, 85 , 86 ,83
11	88 , 87, 86.	85, 84, 87, 85 , 84, 86
12		
13		

14	Sum of observations :	ΣA = 1210	ΣB = 1530
15	Number of observations :	n_A = 14	n_B = 18
16	Means : Σ/n, :	$\bar{A}$ = 86·43	$\bar{B}$ = 85
17	Difference between means, $\bar{A} - \bar{B}$, D = 1·43		
18	Sum of squared observations	ΣA^2 = 104590	ΣB^2 = 130074
19	Correction factors :	$(\Sigma A)^2/n_A$ = 104578·57	$(\Sigma B)^2/n_B$ = 130050
20	Sums-of-squares :	SSA = 11·43	SSB = 24
21	Degrees of freedom : $n-1$:	$d.f._A$ = 13	$d.f._B$ = 17
22	Variance : $SS/d.f.$	V_A = 0·879	V_B = 1·41

23	If variance ratio $\left\{\dfrac{Greater\ V}{Lesser\ V}\right. = \dfrac{1·41}{0·879}$
24	$= 1·6$ $\left.\right\}$ exceeds the
25	F-table value at P ·025, the t test may be unreliable;
26	Use Mann-Whitney (Statform 22) or a transformation
27	Mean variance $= \dfrac{SS_A + SS_B}{d.f._A + d.f._B} = \dfrac{11·43 + 24}{13 + 17}$
28	$= 1·18 = Vm$
29	Standard error $\left.\right\} = \sqrt{Vm\left(\dfrac{1}{n_A} + \dfrac{1}{n_B}\right)} = \sqrt{1·18\left(\dfrac{1}{14} + \dfrac{1}{18}\right)} = 0·386 = SE_D$
30	of difference
31	t-ratio $= \dfrac{\text{Difference } D}{SE_D} = \dfrac{1·43}{0·386}$
32	$= 3·7 = t$-ratio
33	Student's t for d.f. of $(d.f._A + d.f._B)$ at P ·05 is 2·04 $= t$
34	If t-ratio exceeds Student's t, the difference D is "significant"
35	Confidence limits for D are given by $t \times SE_D$, = 0·31 $= C$
36	Confidence range : $\left\{\right.$ Upper : $D + C$ = 1·74
37	$\left.\right\}$ Lower : $D - C$ = 1·12

* R. E. Parker , Introductory Statistics for Biology, Edward Arnold 1973 , p. 22 q 3-4 , p. 92.

ANALYSIS OF VARIANCE FOR SINGLE CLASSIFICATION
UNRESTRICTED RANDOM DESIGN

3 Nature of the observations :

5 Number of treatments, $K =$ .... The number of observations in
6 each treatment, n , need not be constant.
7 Total number of observations, Σn, $N =$

8 Treatments:						Total, Number, Mean, $\downarrow$
10 List						
12 of						
14 observations						
16 y						
18 Sums, $T =$						$= \Sigma y$
19 No., $n =$						$= N$
20 Means, $\bar{y} =$						$= \bar{Y}$
21 $T^2 =$						$\Sigma(T^2/n)$
22 $T^2/n =$						

23 Correction factor $= \dfrac{(\Sigma y)^2}{N} = $ ——————— $= CF = $ _____

25 Treatment sum-of-squares is $\Sigma(T^2/n) - CF,$ $=$ _____
26 Total sum-of-squares, $\Sigma y^2 - CF = $ ....   — $=$ _____
27 Total d.f. $= N-1 = $ .... Treatment d.f. $= K-1$ $= $ ....
28 Residual sum-of-squares and d.f. obtained by subtraction.

29 ANOVAR:	SS	d.f.	MS	V-ratio	Tabulated F
30 Treatments :					(P............)
31 Residual :				$= RMS$	Continue below
32 Total :				if V-ratio exceeds table F.	

33 Follow with Scheffé tests or partitioning (Statforms 8 or 9) or
34 the t-ratio. For the latter, the standard error of a difference
35 between means of two treatments with n and m observations is

36 $SE_D = \sqrt{RMS\left(\frac{1}{n} + \frac{1}{m}\right)} = \sqrt{}$ $=$
37   ....

1	ANALYSIS OF VARIANCE FOR SINGLE CLASSIFICATION
2	UNRESTRICTED RANDOM DESIGN

3 Nature of the observations :

4 **Success-rate of A.I. from six bulls.** *

5 Number of treatments, $K = 6$. The number of observations in
6 each treatment, n, need not be constant.
7 Total number of observations, Σn, $N = 35$

8 Treatments:

	Bull A	B	C	D	E	F	Total, Number, Mean, $\downarrow$
9	46	70	52	47	42	35	
10 (List	31	59	44	21	64	68	
11	37		57	70	50	59	
12 of	62		40	46	69	38	
13	30		67	14	77	57	
14 observations			64		81	76	
15			70		87	57	
16 y						29	
17						60	
18 Sums, $T =$	206	129	394	198	470	479	$1876 = \Sigma y$
19 No., $n =$	5	2	7	5	7	9	$35 = N$
20 Means, $\bar{y} =$	41.2	64.5	56.3	39.6	67.1	53.2	$53.6 = \bar{\bar{Y}}$
21 $T^2 =$	42436	16641	155236	39204	220900	229441	$\Sigma(T^2/n)$
22 $T^2/n =$	8487.2	8320.5	22176.6	7840.8	31557.1	25493.4	103875.6

23
24 Correction factor $= \dfrac{(\Sigma y)^2}{N} = \dfrac{3519376}{35} = CF = 100353.6$

25 Treatment sum-of-squares is $\Sigma(T^2/n) - CF,$ $= 3322$
26 Total sum-of-squares, $\Sigma y^2 - CF = 111076 - 100553.6 = 10522.4$
27 Total d.f. $= N - 1 = 34$ Treatment d.f. $= K - 1 = 5$
28 Residual sum-of-squares and d.f. obtained by subtraction.

29 ANOVAR:	SS	d.f.	MS	V-ratio	Tabulated F
30 Treatments :	3322	5	664.4	2.68 *	2.55 (P.05)
31 Residual :	7200	29	248.3 = RMS	Continue below	
32 Total :	10522	34	if V-ratio exceeds table F.		

33 Follow with Scheffé tests or partitioning (Statforms 8 or 9) or
34 the t-ratio. For the latter, the standard error of a difference
35 between means of two treatments with n and m observations is

36
37 $SE_D = \sqrt{RMS\left(\dfrac{1}{n} + \dfrac{1}{m}\right)} = \sqrt{248\left(\dfrac{1}{5} + \dfrac{1}{9}\right)} = 8.8 \text{ (Bull A vs F)}$

* Snedecor & Cochran, Statistical Methods, Iowa State
University Press 1967, p. 290.

1	SCHEFFÉ CONTRASTS : differences between selected groups of
2	treatments, and their confidence limits.
3	Nature of the observations :
4	

		SS	d.f.	MS	V-ratio	Tabulated F for given P
5	From Anovar, e.g. statforms 7, 10, 11 or 13 D.					
6						(P ....)
7	Treatments :					
8	Residual :				= RMS	

9	Calculation of contrasts and their Scheffé confidence limits :		
10	Labels for each contrast :		

	Treatments			Totals		Totals		Totals	
11				Positive	Negative	Positive	Negative	Positive	Negative
12	Label	Total	No. obs.						
13									
14									
15									
16									
17									
18									
19									
20									
21									
22									

23	Sum of positive totals =		
24	Sum of negative totals =		
25	No of observations, each side, n =		
26	∴ mean on each side =		
27	∴ Difference or Contrast, C =		
28	(Residual MS) ÷ n =		
29	Sum of RMS/n, both sides =		
30	Multiply by treatment d.f. :		
31	Multiply by tabulated F :		
32	Square rooted to give S =		
33	If S exceeds $|C|$ then confidence limits span zero and the contrast		
34	is not significant at the P level selected for table F.		
35	Confidence limits { Upper, $C+S$ =		
36			
37	Lower, $C-S$ =		

SCHEFFÉ CONTRASTS: differences between selected groups of treatments, and their confidence limits.

Nature of the observations:

Further analysis of observations from Statform 7.

From Anovar, e.g. statforms 7, 10, 11 or 13D.

	SS	d.f.	MS	V-ratio	Tabulated F for given P
Treatments:	3322	5	664.4	2.68*	2.55 (P .05 .....)
Residual:	7200	29	248.3 = RMS		

Calculation of contrasts and their Scheffé confidence limits:

Labels for each contrast: Bulls C:D | (C+F):(B+E) | (A+D):(B+E)

Label	Total	No. obs.	C:D Positive	C:D Negative	(C+F):(B+E) Positive	(C+F):(B+E) Negative	(A+D):(B+E) Positive	(A+D):(B+E) Negative
Bull A	206	5					206	
" B	129	2				129		129
" C	394	7		394	394			
" D	198	5					198	
" E	470	7	470			470		470
" F	479	9			479			
Sum of positive totals =			470		873		404	
Sum of negative totals =				394		599		599
No of observations, each side, n =			7	7	16	9	10	9
∴ mean on each side =			67.1	56.3	54.6	66.6	40.4	66.6
∴ Difference or Contrast, C =			10.8		12		26.2	
(Residual MS) ÷ n =			35.5	35.5	15.5	27.6	24.8	27.6
Sum of RMS/n, both sides =			71		43.1		52.4	
Multiply by treatment d.f.:			355		215.4		262	
Multiply by tabulated F:			905		549.4		668	
Square rooted to give S =			30.1		23.4		25.8	

If S exceeds |C| then confidence limits span zero and the contrast is not significant at the P level selected for table F.

Confidence limits	Upper, C+S =	n.s.	n.s.	52
	Lower, C−S =			0.4

1	PARTITIONING TREATMENT SUM−OF−SQUARES;
2	RESPONSE CURVES OR OTHER ORTHOGONAL COMPONENTS
3	(Provided that dose-rates are equally spaced and replicated)

4   Nature of the observations :

5

6   Number of replicates of each treatment level, $n =$ _____

7   From preliminary Anovar, brought forward from analyses on
8   Statforms 7, 10, 11, or 13D :

	Source	SS	d.f.	MS	V-ratio	Table F
9						
10	Treatments:	....		....		
11	Residual :				= R MS	

	Treatment		Coefficients for partitioning : β					
12			($\Sigma \beta$ must be zero in each column)					
13		Totals						
14	Levels	T	Lin.	Quad.	Cub.	____	____	____
15								
16								
17								
18								
19								
20								
21								
22								

23   Calculation of sums−of−squares and variance−ratios :

24	$\Sigma(\beta T)$,	$D =$						
25	$\Sigma(\beta^2)$,	$=$						
26	$\Sigma(\beta^2) \times n$,	$N =$						
27	D^2/N,	$SS =$						
28	SS/RMS,	V-ratio =						

29   V-ratios to be tested against F−table for d.f. 1 over d.f. for residual

30   Calculation of regression coefficient for a linear effect :
31   Interval between adjacent linear β coefficients: $\beta' =$ ....
32   Interval between adjacent dose-rates, $I =$ ....

33
34   Coefficient $b = \dfrac{D \times \beta'}{n \times I} =$ _____ $=$ ....

35   Mean difference between two sides of a comparison :

36
37   Difference $= \dfrac{2 \times D}{\Sigma |\beta| \times n} = \dfrac{2 \times}{\times}$ $=$ ....

1	PARTITIONING TREATMENT SUM−OF−SQUARES;
2	RESPONSE CURVES OR OTHER ORTHOGONAL COMPONENTS
3	(Provided that dose-rates are equally spaced and replicated)

4 Nature of the observations : Detection of Response − curves
5 in observations from Statform 11 , (five spacings).
6 Number of replicates of each treatment level, $n = 5$

7 From preliminary Anovar, brought forward from analyses on
8 Statforms 7, 10, 11, or 13D :

	Source	SS	d.f.	MS	V-ratio	Table F
10	Treatments:	0.370	4	0.093	0.823	3.26
11	Residual :	1.360	12	0.113	= RMS	

	Treatment Levels	Totals T	Coefficients for partitioning : β ($\Sigma \beta$ must be zero in each column)			e.g comparing 8+10 with 2+4
14			Lin.	Quad.	Cub.	
16 (spacings)						
17	2	13.4	−2	2	−1	−1
18	4	13.2	−1	−1	2	−1
19	6	12.6	0	−2	0	0
20	8	12.0	1	−1	−2	1
21	10	11.9	2	2	1	1

23 Calculation of sums−of−squares and variance−ratios :

			Lin.	Quad.	Cub.		comparing
24	$\Sigma(\beta T)$,	$D =$	4.2	0.2	0.9		−2.7
25	$\Sigma(\beta^2)$,	$=$	10	14	10		4
26	$\Sigma(\beta^2) \times n$,	$N =$	50	70	50		20
27	D^2/N,	$SS =$	0.35	0.001	0.016		0.365
28	SS/RMS,	V-ratio =	3.1	0.009	0.142		3.226

29 V-ratios to be tested against F−table for d.f. 1 over d.f. for residual

30 Calculation of regression coefficient for a linear effect :
31 Interval between adjacent linear β coefficients : $\beta' = $ 1
32 Interval between adjacent dose-rates, $I = $ 2 (inches)

34 Coefficient $b = \dfrac{D \times \beta'}{n \times I} = \dfrac{4.2 \times 1}{5 \times 2} = $ 0.42

35 Mean difference between two sides of a comparison : for example only 8+10 : 2+4

37 Difference $= \dfrac{2 \times D}{\Sigma |\beta| \times n} = \dfrac{2 \times -2.7}{4 \times 5} = $ 0.27

STATFORM 10

1	ANALYSIS OF VARIANCE FOR RANDOMISED BLOCK
2	or any two-way classification without interactions.
3	Nature of the observations, y :
4	
5	Number of treatments, K = ...
6	Number of blocks, n = ...
7	Total number of observations, $K \times n$, N =

8	Table of observations :							Treatment	
9	Block No :	1	2	3	4	5	6	Totals T	Means
10	Treatments								
11									
12									
13									
14									
15									
16									
17									
18									
19	Block { Totals, B :								= Σy
20	Means :							$\bar{Y}$ =	

21	Correction factor, $(\Sigma y)^2/N$ = ... / = ... = CF
22	Total sum-of-squares, $\Sigma(y^2) - CF$ = ... − = ...
23	Treatment SS = $\dfrac{\Sigma(T^2)}{n} - CF$ = —— − =
24	= ....
25	Block SS = $\dfrac{\Sigma(B^2)}{K} - CF$ = —— − =
26	= ....
27	Treatment d.f. = $K-1$. Block d.f. = $n-1$. Total d.f. = $N-1$.

28	ANOVAR :	SS	d.f.	MS	V-ratio	Table F
29	Blocks :					}
30	Treatments :					
31	Residual :				Residual SS and d.f.	
32	Total :				obtained by subtraction	

33	Standard error of difference between two treatment means :
34	$SE_d = \sqrt{\dfrac{2 \times RMS}{n}} = \sqrt{\dfrac{2 \times}{}}$ =
35	....
36	Differences between means to be examined by Scheffé test or partitioning
37	(Statforms 8 or 9) or t-ratio according to nature of the design.

STATFORM 10

1	ANALYSIS OF VARIANCE FOR RANDOMISED BLOCK
2	or any two-way classification without interactions.
3	Nature of the observations, y : Failure to emerge, out of
4	100 seed. (n.b normally should have angular transformation). *
5	Number of treatments, K = 5
6	Number of blocks, n = 5
7	Total number of observations, K × n, N = 25

8 Table of observations :

Block No:	1	2	3	4	5	6	Treatment Totals T	Means
Treatments								
Check	8	10	12	13	11		54	10·8
Arasan	2	6	7	11	5		31	6·2
Spergon	4	10	9	8	10		41	8·2
Semesan	3	5	9	10	6		33	6·6
Fermate	9	7	5	5	3		29	5·8
Block {Totals, B :	26	38	42	47	35		188 = Σy	
Block {Means :	5·2	7·6	8·4	9·4	7·0		Ȳ = 7·52	

21	Correction factor, $(\Sigma y)^2/N$ = 35344/25 = 1413·76 = CF
22	Total sum-of-squares, $\Sigma(y^2) - CF$ = 1634 − 1413·76 = 220·24
23-24	Treatment SS = $\dfrac{\Sigma(T^2)}{n} - CF = \dfrac{7488}{5} - 1413·76 = 83·84$
25-26	Block SS = $\dfrac{\Sigma(B^2)}{K} - CF = \dfrac{7318}{5} - 1413·76 = 49·84$
27	Treatment d.f. = K−1. Block d.f. = n−1. Total d.f. = N−1.

ANOVAR :	SS	d.f.	MS	V-ratio	Table F
Blocks :	49·84	4	12·46	2·3	3·01 }
Treatments :	83·84	4	20·96	3·87	3·01 } P.05
Residual :	86·56	16	5·41	Residual SS and d.f.	
Total :	220·24	24		obtained by subtraction	

33	Standard error of difference between two treatment means :
34-35	$SE_d = \sqrt{\dfrac{2 \times RMS}{n}} = \sqrt{\dfrac{2 \times 5·41}{5}} = 1·47$
36	Differences between means to be examined by Scheffé test or partitioning
37	(Statforms 8 or 9) or t-ratio according to nature of the design.

* Snedecor & Cochran, Statistical Methods, Iowa State
University Press 1967, p. 300.

ANALYSIS OF VARIANCE FOR LATIN SQUARE
(up to 5 × 5)

Nature of the observations :

No. of treatments, n = ... Total no. of observations, N = ...

T = treatment totals. C = column totals. R = row totals.

Layout plan Column numbers					Row		Treatment table Treatment labels				
1	2	3	4	5	No.	Totals	A	B	C	D	E
					1						
					2						
					3						
					4						
					5						
					Totals :						
Column totals					Means :						

Correction factor $\Big\} = \dfrac{(\sum y)^2}{N}$ = ————— = ... $= CF$

Calculation of sums-of-squares :

Rows	Columns	Treatments	Total
$\sum (R^2/n)$ = ...	$\sum (C^2/n)$ =	$\sum (T^2/n)$ =	$\sum (y^2)$ =
CF = ————	CF = ————	CF = ————	CF = ————
SS_R =	SS_C =	SS_T =	SS_y =

ANOVAR :

	SS	d.f.	MS	V-ratio	Table F
Rows :					
Columns :					$\Big\}$ (P. ......)
Treatments :					
Residual :	————	————		Residual SS and	
Total :				d.f. obtained by subtraction	

Standard error of a difference between two treatment means :

$$SE_d = \sqrt{\frac{2 \times RMS}{n}} = \sqrt{\frac{2 \times \rule{2cm}{0.4pt}}{}} = \text{............}$$

Differences between means to be examined by Scheffé test or partitioning (Statforms 8 or 9) or t-ratio according to nature of the design

STATFORM 11

1	ANALYSIS OF VARIANCE FOR LATIN SQUARE
2	(up to 5 × 5)

3 *Nature of the observations :*

4 Millet yield in hundreds of grams at five spacings. *

5 No. of treatments, n = 5   Total no. of observations, N = 25

6 T = treatment totals. C = column totals. R = row totals.

Layout plan — Column numbers

1	2	3	4	5	Row No.	Totals	A	B	C	D	E
B 2.6	E 2.3	A 2.8	C 2.9	D 2.0	1	12.6	2.8	2.6	2.9	2.0	2.3
D 2.5	A 2.8	E 2.5	B 2.8	C 2.6	2	13.2	2.8	2.8	2.6	2.5	2.5
E 1.8	B 2.5	C 2.8	D 2.5	A 2.5	3	12.1	2.5	2.5	2.8	2.5	1.8
A 2.0	C 2.0	D 2.3	E 1.9	B 2.6	4	10.8	2.0	2.6	2.0	2.3	1.9
C 2.3	D 2.7	B 2.7	A 3.3	E 3.4	5	14.4	3.3	2.7	2.3	2.7	3.4
11.2	12.3	13.1	13.4	13.1	Totals:		13.4	13.2	12.6	12.0	11.9
	Column totals				Means:		2.68	2.64	2.52	2.40	2.38

Treatment table — Treatment labels (A B C D E)

17
18
$$\text{Correction factor} = \frac{(\Sigma y)^2}{N} = \frac{63.1^2}{25} = 159.264 = CF$$

19 Calculation of sums-of-squares :

	Rows	Columns	Treatments	Total
20				
21	$\Sigma(R^2/n) = 160.682$	$\Sigma(C^2/n) = 159.902$	$\Sigma(T^2/n) = 159.634$	$\Sigma(y^2) = 163.050$
22	$CF = \underline{159.264}$	$CF = \underline{159.264}$	$CF = \underline{159.264}$	$CF = \underline{159.264}$
23	$SS_R = 1.418$	$SS_C = 0.638$	$SS_T = 0.370$	$SS_y = 3.786$

24 ANOVAR :	SS	d.f.	MS	V-ratio	Table F
25 Rows :	1.418	4	0.355	3.1	
26 Columns :	0.638	4	0.160	1.4	} 3.26 (P.05)
27 Treatments :	0.370	4	0.093	<1	
28 Residual :	1.360	12	0.113	Residual SS and	
29 Total :	3.786	24		d.f. obtained by subtraction	

30 Standard error of a difference between two treatment means :

31
32
$$SE_d = \sqrt{\frac{2 \times RMS}{n}} = \sqrt{\frac{2 \times 0.113}{5}} = 0.213$$

33 Differences between means to be examined by Scheffé test or partitioning
34 (Statforms 8 or 9) or t-ratio according to nature of the design

* Snedecor & Cochran, Statistical Methods, Iowa State University Press 1967, p. 313.

STATFORM 12A

1	ANALYSIS OF VARIANCE FOR 2^3 or 2^2 FACTORIAL EXPERIMENT
2	Nature of the observations :
3	Description ⎧ a :
4	of the ⎨ b :
5	factors ⎩ c :
6	No. of factors (2 or 3) n = ... No. of replicates or blocks, r = ...
7	No. treatments, 2^n, K = No. of observations, K × r, N =

Replicate No.	Observations listed in columns by treatments								Replicate totals
	nil	a	b	ab	c	ac	bc	abc	
10									
11									
12									
13									
14									
15									
16									
17									
18 Totals:									
19 Means:									Σy

20 Signs (coefficients) for calculating treatment effects :

21 Effect	nil	a	b	ab	c	ac	bc	abc
22 A	−	+	−	+	−	+	−	+
23 B	−	−	+	+	−	−	+	+
24 AB	+	−	−	+	+	−	−	+
25 C	−	−	−	−	+	+	+	+
26 AC	+	−	+	−	−	+	−	+
27 BC	+	+	−	−	−	−	+	+
28 ABC	−	+	+	−	+	−	−	+

29 Calculation of sums-of-squares : accumulate sums of negative totals
30 and positive totals from each effect line above.

31 Effect :	A	B	AB	C	AC	BC	ABC
32 Positive totals :							
33 Negative total :							
34 Difference, D =							
35 D^2							
36 D^2/N = SS =							

37 Carry sums-of-squares and means over to Statform 12B.

STATFORM 12A

1	ANALYSIS OF VARIANCE FOR 2^3 or 2^2 FACTORIAL EXPERIMENT
2	Nature of the observations :
3	Description ⎰ a : Boar and Sow
4	of the ⎨ b : Protein 12 & 14 %
5	factors ⎱ c : Lysine zero and 0·6%

Daily gain in weight of 64 pigs. *

6	No. of factors (2 or 3) $n = $ **3**	No. of replicates or blocks, $r = $ **8**	
7	No. treatments, 2^n, $K = $ **8**	No. of observations, $K \times r$, $N = $ **64**	

8 Replicate	Observations listed in columns by treatments								Replicate
9 No.	nil	a	b	ab	c	ac	bc	abc	totals
10 1	1·11	1·03	1·52	1·48	1·22	0·87	1·38	1·09	9·70
11 2	0·97	0·97	1·45	1·22	1·13	1·00	1·08	1·09	8·91
12 3	1·09	0·99	1·27	1·53	1·34	1·16	1·40	1·47	10·25
13 4	0·99	0·99	1·22	1·19	1·41	1·29	1·21	1·43	9·73
14 5	0·85	0·99	1·67	1·16	1·34	1·00	1·46	1·24	9·71
15 6	1·21	1·21	1·24	1·57	1·19	1·14	1·39	1·17	10·12
16 7	1·29	1·19	1·34	1·13	1·25	1·36	1·17	1·01	9·74
17 8	0·96	1·24	1·32	1·43	1·32	1·32	1·21	1·13	9·93
18 Totals:	8·47	8·61	11·03	10·71	10·20	9·14	10·30	9·63	78·09
19 Means:	1·06	1·08	1·38	1·34	1·28	1·14	1·29	1·20	Σy

20 Signs (coefficients) for calculating treatment effects :

21 Effect	nil	a	b	ab	c	ac	bc	abc
22 A	−	+	−	+	−	+	−	+
23 B	−	−	+	+	−	−	+	+
24 AB	+	−	−	+	+	−	−	+
25 C	−	−	−	−	+	+	+	+
26 AC	+	−	+	−	−	+	−	+
27 BC	+	+	−	−	−	−	+	+
28 ABC	−	+	+	−	+	−	−	+

29 Calculation of sums-of-squares: accumulate sums of negative totals
30 and positive totals from each effect line above.

31 Effect :	A	B	AB	C	AC	BC	ABC
32 Positive totals :	38·09	41·67	39·01	39·27	38·27	37·01	39·47
33 Negative total :	40·00	36·42	39·08	38·82	39·82	41·08	38·62
34 Difference, $D = $	−1·91	5·25	−0·07	0·45	−1·55	−4·07	0·85
35 D^2	3·65	27·56	0·005	0·203	2·403	16·56	0·723
36 $D^2/N = SS = $	0·057	0·431	0·0000…	0·003	0·038	0·259	0·011

37 Carry sums-of-squares and means over to Statform 12 B.

* Snedecor & Cochran, Statistical Methods, Iowa State University Press 1967, p. 359 .

STATFORM 12B

2^n FACTORIAL EXPERIMENT: ANALYSIS OF VARIANCE

Continued from Statform 12 A

Correction factor $= \dfrac{(\Sigma y)^2}{N} - $ _____ $= $ .... $= CF$

Total sum-of-squares $= \Sigma y^2 - CF = $ .... $- $ .... $= $ .... $= SSy$

If experiment was disposed in randomized blocks :

Block SS $= \dfrac{\Sigma(\text{replic. total}^2)}{K} - CF = $ _____ $- $ .... $= $ ....

Source	SS	d.f.	MS	V ratio	Table F
Blocks :	....				
A	....	1			
B	....	1			
AB	....	1			
C	....	1			
AC	....	1			
BC	....	1			
ABC	....	1			
Residual :				Residual d.f. and SS	
Total, SSy :				obtained by subtraction.	

For the significant or interesting interactions, construct interaction
tables using the totals or means from Statform 12 A.

AB		− b	+ b		AC		− c	+ c		BC		− c	+ c
		nil + c	b + bc				nil + b	c + bc				nil + a	c + ac
a	−				a	−				b	−		
		a + ac	ab + abc				a + ab	ac + abc				b + ab	bc + abc
	+					+					+		

For the 3-factor interaction ABC, compare :

AB without C :   with: AB with C:

		− b	+ b				− b	+ b
		nil	b				ċ	bc
a	−				a	−		
		a	ab			−	ac	abc
	+							

STATFORM 12B

2^n FACTORIAL EXPERIMENT: ANALYSIS OF VARIANCE

Continued from Statform 12 A

$$\text{Correction factor} = \frac{(\Sigma y)^2}{N} = \frac{6098 \cdot 048}{64} = 95 \cdot 282\ldots = CF$$

$$\text{Total sum-of-squares} = \Sigma y^2 - CF = 97 \cdot 321 - 95 \cdot 282 = 2 \cdot 039 = SSy$$

If experiment was disposed in randomized blocks :

$$\text{Block SS} = \frac{\Sigma (\text{replic. total}^2)}{K} - CF = \frac{763 \cdot 385}{8} - 95 \cdot 282 = 0 \cdot 141\ldots$$

Source	SS	d.f.	MS	V ratio	Table F
Blocks :	0·141	7	0·020	<1	2·2
A	0·057	1	0·057	2·5	
B	0·431	1	0·431	19·2***	
AB	0·000	1	0·000	<1	
C	0·003	1	0·003	<1	4·0
AC	0·038	1	0·038	1·7	
BC	0·259	1	0·259	11·5***	
ABC	0·011	1	0·011	<1	
Residual :	1·099	49	0·0224	Residual d.f. and SS	
Total, SSy :	2·039	63	obtained by subtraction.		

For the significant or interesting interactions, construct interaction tables using the totals or means from Statform 12 A.

AB		b −	b +		AC		c −	c +		BC		c −	c +
		nil + c	b + bc				nil + b	c + bc				nil + a	c + ac
a −		18·67	21·33		a −		19·5	20·5		b −		17·08	19·34
		a + ac	ab + abc				a + ab	ac + abc				b + ab	bc + abc
a +		17·75	20·34		a +		19·32	18·77		b +		21·74	19·93

For the 3-factor interaction ABC, compare : non-sig but entered
AB without C : with: AB with C : for example only

		b −	b +				b −	b +
		nil	b				c	bc
a −		8·5	11·0		a −		10·2	10·3
		a	ab				ac	abc
a +		8·6	10·7				9·1	9·6

STATFORM 13A

1	ANALYSIS OF VARIANCE FOR A FACTORIAL WITH MORE
2	THAN TWO LEVELS AND UP TO FOUR FACTORS.
3	_Stage 1 : arrangement of data and calculation of totals._
4	Nature of the observations :
5	

Factor Label	No. of Levels	Description	Factor Label	No. of Levels	Description
A :			C :		
B :			D :		

9 (Include blocks or replicates as a "factor" at this stage)

10 Arrange observations in a single column (OBS) in hierarchical sequence,
11 incrementing last factors first.

Factor Levels				OBS	TOTALS			Factor Levels				OBS	TOTALS		
A	B	C	D		C	B	A	A	B	C	D		C	B	A
14															
15															
16															
17															
18															
19															
20															
21															
22															
23															
24															
25															
26															
27															
28															
29															
30															
31															
32															
33															
34															

35 Sum of observations, Σy = ...   Sum of squared observations, Σy^2 = ...

36 No. of observations, N =   Correction factor, $(\Sigma y)^2/N$ = $= CF$

37 Continue on Statform 13B

STATFORM 13A

1. ANALYSIS OF VARIANCE FOR A FACTORIAL WITH MORE
2. THAN TWO LEVELS AND UP TO FOUR FACTORS.
3. *Stage 1 : arrangement of data and calculation of totals.*
4. Nature of the observations :
5. **Gain in weight of 48 pigs on 3 diet supplements.** *

Factor Label	No. of Levels	Description	Factor Label	No. of Levels	Description
A :	4	Lysine	C :	2	Protein
B :	3	Methionine	D :	2	Blocks

9. (Include blocks or replicates as a "factor" at this stage)
10. Arrange observations in a single column (OBS) in hierarchical sequence,
11. incrementing last factors first.

A	B	C	D	OBS	C	B	A	A	B	C	D	OBS	C	B	A
1	1	1	1	1.11				3	1	1	1	1.22			
			2	0.97	2.08						2	1.13	2.35		
		2	1	1.52						2	1	1.38			
			2	1.45	2.97	5.05					2	1.08	2.46	4.81	
	2	1	1	1.09					2	1	1	1.34			
			2	0.99	2.08						2	1.41	2.75		
		2	1	1.27						2	1	1.40			
			2	1.22	2.49	4.57					2	1.21	2.61	5.36	
	3	1	1	0.85					3	1	1	1.34			
			2	1.21	2.06						2	1.19	2.53		
		2	1	1.67						2	1	1.46			
			2	1.24	2.91	4.97	14.59				2	1.39	2.85	5.38	15.55
2	1	1	1	1.30				4	1	1	1	1.19			
			2	1.00	2.30						2	1.03	2.22		
		2	1	1.55						2	1	0.80			
			2	1.53	3.08	5.38					2	1.29	2.09	4.31	
	2	1	1	1.03					2	1	1	1.36			
			2	1.21	2.24						2	1.16	2.52		
		2	1	1.24						2	1	1.42			
			2	1.34	2.58	4.82					2	1.39	2.81	5.33	
	3	1	1	1.12					3	1	1	1.46			
			2	0.96	2.08						2	1.03	2.49		
		2	1	1.76						2	1	1.62			
			2	1.27	3.03	5.11	15.31				2	1.27	2.89	5.38	15.02

35. Sum of observations, $\Sigma y = 60.47$ | Sum of squared observations, $\Sigma y^2 = 78.2205$
36. No. of observations, $N = 48$ | Correction factor, $(\Sigma y)^2 / N = 76.1796 = CF$
37. Continue on Statform 13B

* Snedecor & Cochran, Statistical Methods, Iowa State University Press 1967, p.362.

STATFORM 13B

1	FACTORIAL ANALYSIS OF VARIANCE Continued from Statform 13A
2	Stage 2: 2-factor interaction tables and sums-of-squares.
3	Interaction tables: totals from Statform 13A.

From Statform 13A: $\Sigma y^2 = $ ....
Less CF, $(\Sigma y)^2/N$ = _____
gives total sum-of-squares:

Matrix of
maineffect and
interaction
totals

18	Check: interaction totals must add up to main effect totals in the diagonal.
19	Calculation of sums-of-squares; n always indicates the number of
20	observations contributing to each squared total.

		A	B	C	D
21	Main-effect:				
22	$\Sigma(T^2)$:				
23	n:				
24	$\Sigma(T^2)/n$:				
25	Less CF:				
26	Maineffect SS:				

		AB	AC	AD	BC	BD	CD
27	Interaction:						
28	$\Sigma(T^2)$:						
29	n:						
30	$\Sigma(T^2)/n$:						
31	Less CF:						
32	Subtotal SS:						
33	Less SS_A:						
34	" SS_B:						
35	" SS_C:						
36	" SS_D:						
37	Interaction SS:						

STATFORM 13B

1	FACTORIAL ANALYSIS OF VARIANCE *Continued from Statform 13A*
2	Stage 2 : 2-factor interaction tables and sums-of-squares.
3	Interaction tables : totals from Statform 13A.

	A 1	A 2	A 3	A 4	B 1	B 2	B 3	C 1	C 2	(D not a factor)
A 1	14.59									
2		15.31								
3			15.55							
4				15.02						
B 1	5.05	5.38	4.81	4.31	19.55					
2	4.57	4.82	5.36	5.33		20.08				
3	4.97	5.11	5.38	5.38			20.84			
C 1	6.22	6.62	7.63	7.23	8.95	9.59	9.16	27.70		
2	8.37	8.69	7.92	7.79	10.60	10.49	11.68		32.77	
D 1	(not a factor)								31.50	
2										28.97

From Statform 13A: $\Sigma y^2 = 78.2205$
Less CF, $(\Sigma y)^2/N$ = 76.1796
gives total sum-of-squares: 2.0409

Matrix of maineffect and interaction totals

18	Check: *interaction totals must add up to main effect totals in the diagonal.*
19	Calculation of sums-of-squares; n always indicates the number of
20	observations contributing to each squared total.

Main-effect :	A	B	C	D
$\Sigma(T^2)$:	914.6671	1219.7145	1841.1629	1831.5109
n :	12	16	24	24
$\Sigma(T^2)/n$:	76.2223	76.2322	76.7151	76.3130
Less CF :	76.1796	76.1796	76.1796	76.1796
Maineffect SS :	0.0427	0.0526	0.5355	0.1334

Interaction :	AB	AC	AD	BC	BD	CD
$\Sigma(T^2)$:	306.1167	461.9861		614.7987		
n :	4	6		8		
$\Sigma(T^2)/n$:	76.5292	76.9977		76.8498		
Less CF :	76.1796	76.1796		76.1796		
Subtotal SS :	0.3496	0.8181		0.6702		
Less SSA :	0.0427	0.0427				
" SSB :	0.0526			0.0526		
" SSC :		0.5355		0.5355		
" SSD :						
Interaction SS :	0.2543	0.2399		0.0821		

STATFORM 13C

1	FACTORIAL ANALYSIS OF VARIANCE *Continued from Statform 13B.*
2	Stage 3 : 3-factor interactions.
3	Interaction tables: Construct 3-dimensional tables, using totals taken
4	from Statform 13A, listed in rows and columns labelled by factors.
5	The number (n) of observations contributing to each total must
6	also be known.
7	
8	
9	
10	
11	
12	
13	
14	
15	
16	
17	
18	
19	
20	
21	

		ABC	ABD	ACD	BCD
22	Calculation of sums-of-squares. Main-effect and 2-factor				
23	sums-of-squares and correction factors taken from Statform 13B.				
24	Interaction :				
25	$\Sigma (T^2)$:				
26 27	$\Sigma (T^2)/n$:				
28	Less CF :				
29	Subtotal SS :				
30	Less SS for :	A :			
31	"	B :			
32	"	C :			
33	"	AB :			
34	"	AC :			
35	"	BC :			
36	Interaction SS :				
37	Continue on Statform 13 D				

1	FACTORIAL ANALYSIS OF VARIANCE *Continued from Statform 13B*
2	Stage 3 : 3-factor interactions.
3	Interaction tables: Construct 3-dimensional tables, using totals taken
4	from Statform 13A, listed in rows and columns labelled by factors.
5	The number (n) of observations contributing to each total must
6	also be known.

	MET B	PROT C	LYS A_1	A_2	A_3	A_4
9	1	1	2.08	2.30	2.35	2.22
10		2	2.97	3.08	2.46	2.09
11	2	1	2.08	2.24	2.75	2.52
12		2	2.49	2.58	2.61	2.81
13	3	1	2.06	2.08	2.53	2.49
14		2	2.91	3.03	2.85	2.89

		ABC	ABD	ACD	BCD
22	Calculation of sums-of-squares. Main-effect and 2-factor				
23	sums-of-squares and correction factors taken from Statform 13B.				
24	Interaction :	ABC	ABD	ACD	BCD
25	$\Sigma (T^2)$:	154.9105			
26	n :	2			
27	$\Sigma (T^2)/n$:	77.4553			
28	Less CF :	76.1796			
29	Subtotal SS :	1.2757			
30	Less SS for :	A : 0.0427			
31	"	B : 0.0526			
32	"	C : 0.5355			
33	"	AB : 0.2543			
34	"	AC : 0.2399			
35	"	BC : 0.0821			
36	Interaction SS :	0.0686			
37	Continue on Statform 13D				

1	FACTORIAL ANALYSIS OF VARIANCE Continued from Statform 13C
2	Stage 4 : Anovar table, designed to accommodate two and three-factor
3	interactions and subtotals if for a split-plot. For simpler designs omit
4	irrelevant lines

	Treatments		
5			
6	Symbol	Levels	Description
7	A		....
8	B		....
9	C		....
10	D		

11	Main-effect and 2-factor interaction sums—of—squares from
12	Statform 13B, 3-factor interactions and subtotals (if required)
13	from Statform 13C, enter in Anovar table:

	ANOVAR : Source	SS	d.f.	MS	Variance ratio	Table F
14						
15						
16	A					
17	B					
18	AB or residual					
19	Subtotal					
20						
21	C					
22	AC					
23	BC					
24	ABC or residual					
25	Subtotal					
26						
27	D					
28	AD					
29	BD					
30	ABD					
31	CD					
32	ACD					
33	BCD					
34	Residual				Residual d.f. and	
35	Grand total			S.S. obtained by subraction		

36	Effects with sufficiently high variance ratios may be further examined
37	as in Statforms 8, 9 or 10, as suitable.

STATFORM 13D

1	FACTORIAL ANALYSIS OF VARIANCE *Continued from Statform 13C*						
2	*Stage 4 : Anovar table, designed to accommodate two and three-factor*						
3	*interactions and subtotals if for a split-plot. For simpler designs omit*						
4	*irrelevant lines*						

5 Treatments

	Symbol	Levels	Description			
7	A	4	Lysine			
8	B	3	Methionine	} supplements to diet.		
9	C	2	Protein			
10	D	2	Replicates = Blocks			

11 Main-effect and 2-factor interaction sums—of—squares from
12 Statform 13B, 3-factor interactions and subtotals (if required)
13 from Statform 13C, enter in Anovar table: (P.05)

14 ANOVAR :

	Source		SS	d.f.	MS	Variance ratio	Table F
16	A	Lysine	0·0427	3	0·0142	<1	
17	B	Methionine	0·0526	2	0·0263	<1	
18	AB or residual	LxM	0·2543	6	0·0424	1·5	2·53
19	Subtotal		—				
20							
21	C	Protein	0·5355	1	0·5355	19·5 ***	4·28
22	AC	LxP	0·2399	3	0·0800	2·9	3·03
23	BC	MxP	0·0821	2	0·0411	1·49	3·42
24	ABC or residual	LxMxP	0·0686	6	0·0114	<1	
25	Subtotal		—				
26							
27	D	Blocks	0·1334	1	0·1334	4·85 *	4·28
28	AD		—				
29	BD		—				
30	ABD		—				
31	CD		—				
32	ACD		—				
33	BCD		—				
34	Residual		0·6318	23	0·0275	Residual d.f. and	
35	Grand total		2·0409	47		S.S. obtained by subtraction	

36 Effects with sufficiently high variance ratios may be further examined
37 as in Statforms 8, 9 or 10, as suitable.

LINEAR REGRESSION AND CORRELATION
for a small set of observations

Nature of the observations :

Dependent or LH variable y : | Independent or RH variable x :

No. of observations of each, i.e. no. of pairs, n =

Calculation of sums of squares and products :

	Observations	
	y	x

$(\Sigma y)^2$ = ___ | $\Sigma y \times \Sigma x$ = ___ | $(\Sigma x)^2$ = ___

$\Sigma(y^2)$ = ___ | $\Sigma(yx)$ = ___ | $\Sigma(x^2)$ = ___

$(\Sigma y)^2/n$ = ___ | $\Sigma y \Sigma x/n$ = ___ | $(\Sigma x)^2/n$ = ___

SS_y = ___ | SP_{yx} = ___ | SS_x = ___

Regression coefficient } "Slope" = $\dfrac{SP_{yx}}{SS_x}$ = ——— = ___ = b

Regression constant } "intercept" = $\bar{y} - b \times \bar{x}$ = ___ = a

Regression sum-of-squares } = $b \times SP_{yx}$ = ___

ANOVAR : Total sum-of-squares is SS_y. Total d.f. is $n-1$. Residual SS and d.f. obtained by subtraction of regression values from the total.

Source	SS	d.f.	MS	Variance ratio	Table F
Regression :		1			
Residual :					(P.___)
Total :					

Continue if variance ratio exceeds table F.

Coefficient of determination } $r^2 = \dfrac{\text{Regression SS}}{\text{Total SS}}$ = ——— = ___

Correlation coefficient } $r = \sqrt{r^2}$ = ___ { with sign as for b

Standard error of coefficient b } = $\sqrt{\dfrac{\text{Residual MS}}{SS_x}}$ = $\sqrt{\dfrac{}{}}$ = ___

Standard error of mean of random sample of y estimated from n values of x } = $\sqrt{\dfrac{\text{Residual MS}}{n}}$ = ___

For confidence limits continue to Statform 15.

	Σy	Σx
	$\bar{y}$	$\bar{x}$
	Σy^2	Σx^2
	$\Sigma(yx)$	

STATFORM 14

1	LINEAR REGRESSION AND CORRELATION
2	*for a small set of observations*

3 Nature of the observations : units , mg./100mL.. *

4 Dependent or LH variable y : | Independent or RH variable x :

5 Cholesterol in blood serum | Age of eleven women.

6 of Iowa women.

7 No. of observations of each, i.e. no. of pairs, n = 11

8 Calculation of sums of squares and products :

			Observations	
			y	x
9	$(\Sigma y)^2$ = **519.84** $\Sigma y \times \Sigma x$ = **13315.2** $(\Sigma x)^2$ = **341056**			
10	$\Sigma(y^2)$ = **51.44** $\Sigma(yx)$ = **1270.2** $\Sigma(x^2)$ = **32834**		1.8	46
11	$(\Sigma y)^2/n$ = **47.26** $\Sigma y \Sigma x/n$ = **1210.47** $(\Sigma x)^2/n$ = **31005.1**		2.3	52
12	SSy = **4.18** $SPyx$ = **59.73** SSx = **1828.9**		1.8	39

13 Regression ⎰ "Slope" = $\dfrac{SPyx}{SSx}$ = $\dfrac{59.73}{1828.9}$ = **0.03266** = b | 2.5 | 65

14 coefficient ⎱

15 Regression ⎰ "intercept" = $\bar{y} - b \times \bar{x}$ = **0.336** = a | 2.6 | 54

16 constant ⎱ | 2.0 | 33

17 Regression ⎰ = $b \times SPyx$ = **1.95** | 1.2 | 49

18 sum-of-squares ⎱ | 3.4 | 76

19 ANOVAR : Total sum-of-squares is SSy. Total d.f. | 2.2 | 71

20 is n−1. Residual SS and d.f. obtained by subtraction | 1.1 | 41

21 of regression values from the total. | 1.9 | 58

22 23	Source	SS	d.f.	MS	Variance ratio	Table F
24	Regression :	1.95	1	1.95	7.86	5.12
25	Residual :	2.23	9	0.248		(P.05)
26	Total :	4.18	10			

27 Continue if variance ratio exceeds table F.

28 Coefficient of ⎰ $r^2 = \dfrac{\text{Regression SS}}{\text{Total SS}}$ = $\dfrac{1.95}{4.18}$ = **0.467**

29 determination ⎱

30 Correlation ⎰ $r = \sqrt{r^2}$ = **0.68** ⎰ with sign

31 coefficient ⎱ ⎱ as for b

32 Standard error of ⎰ = $\sqrt{\dfrac{\text{Residual MS}}{SSx}}$ = $\sqrt{\dfrac{0.248}{1828.9}}$ = **0.0116**

33 coefficient b ⎱

34 Standard error of mean of random sample of y

35 estimated from ⎰ = $\sqrt{\dfrac{\text{Residual MS}}{n}}$ = **0.150**

36 n values of x ⎱

37 For confidence limits continue to Statform 15.

Σy	Σx
22.8	5.84
$\bar{y}$	$\bar{x}$
2.07	53.1
Σy^2	Σx^2
51.44	32834
$\Sigma(yx)$	
1270.2	

* Snedecor & Cochran , Statistical Methods, Iowa State University Press 1967 , p.433

1	CONFIDENCE LIMITS FOR LINEAR REGRESSION
2	following from Statform 14.

3	Nature of the observations :
4	Dependent or LH variable y : Independent or RH variable x :
5	
6	

7	Required from previous calculations, Statform 14 :
8	Mean of x, $\bar{x}$ = ... Regression constant, a = ...
9	No. of observations, n = ... Regression coefficient, b = ...
10	Student's t for $n-2$ degrees of Sum of squares of x, SSx = ...
11	freedom and P. .... is: t = Residual mean square, RMS =

12	Nature of required confidence limits defined by m, which is the size of
13	the population for which the limits are required :
14	(1) for single observations, $m = 1$.
15	(2) for means of several observations, m is set to the
16	required number.
17	(3) for unlimited observations or for the regression line itself,
18	m is taken as infinity and $1/m$ below is taken as zero.

19	Calculation of the confidence limits :
20	Points on the x-axis at which confidence limits for y are required
21	are designated below by x'.

22	x' Values :				
23	$(x' - \bar{x})$:				
24	Squared :				
25	Divided by SSx :				
26	Add $1/n$ (=) :				
27	Add $1/m$ (=) :				
28	Multiply by RMS :				
29	Square-rooted to } :				
30	give standard error)				
31	Multiplied by t } c =				
32	to give				
33	Upper confidence } $y + c$ =				
34	limit				
35	Estimate of y : $(b \times x') + a$ =				
36	Lower confidence } $y - c$ =				
37	limit				

STATFORM 15

#						
1	CONFIDENCE LIMITS FOR LINEAR REGRESSION					
2	*following from Statform 14.*					

3 Nature of the observations : observations from Statform 14.

4 Dependent or LH variable y: Independent or RH variable x:

5 cholesterol in blood   Age of women.

6 serum of women .

7 Required from previous calculations, Statform 14:

8 Mean of x, $\bar{x}$ = 53·1   Regression constant, a = 0·336

9 No. of observations, n = 11   Regression coefficient, b = 0·0327

10 Student's t for $n-2$ degrees of   Sum of squares of x, SSx = 1829

11 freedom and P .05 is: t = 2·26   Residual mean square, RMS = 0·248

12 Nature of required confidence limits defined by m, which is the size of

13 the population for which the limits are required :

14 (1) for single observations, $m = 1$.

15 (2) for means of several observations, m is set to the

16 required number.

17 (3) for unlimited observations or for the regression line itself,

18 m is taken as infinity and $1/m$ below is taken as zero.

19 Calculation of the confidence limits :

20 Points on the x-axis at which confidence limits for y are required

21 are designated below by x'.

	x' Values :	30	40	50	60	70
23	$(x' - \bar{x})$:	-23·1		3·1		16·9
24	Squared :	533·6		9·6		285·6
25	Divided by SSx :	0·292		0·0052		0·156
26	Add $1/n$ (=) :	0·383		0·0961		0·247
27	Add $1/m$ (=) :	0·383		0·0961		0·247
28	Multiply by RMS :	0·095		0·0238		0·0613
29	Square-rooted to $\Big\}$:					
30	give standard error	0·308		0·154		0·247
31	Multiplied by t $\Big\}$ $c =$					
32	to give	0·696		0·348		0·558
33	Upper confidence $\Big\}$ $y+c =$					
34	limit	2·01		2·32		3·18
35	Estimate of y : $(b \times x') + a =$	1·317		1·971		2·625
36	Lower confidence $\Big\}$ $y-c$ =					
37	limit	0·62		1·62		2·07

COMPARISON OF TWO PROPORTIONS
based on "large" samples, preferably with >30 observations

Nature of the observations :

Group A :

Group B :

Test of difference between the two groups :

	Group A	Group B	Total
No. with the attribute :	$=ca$	$=cb$	$=C$
No. without attribute :	$=da$	$=db$	$=D$
Total no. of observations :	$=A$	$=B$	$=T$

Calculate χ^2 from

$$\frac{T\,(|(ca \times db) - (cb \times da)| - T/2)^2}{A \times B \times C \times D} \quad \text{as below}$$

$ca \times db =$	$A =$	
$cb \times da =$ _____	Multiply by B :	
Subtract and make positive	Multiply by C :	
$0.5 \times T$ _____	Multiply by D : _____ $=(2)$	
Subtract :		
Square :	$\chi^2 = \dfrac{(1)}{(2)} =$	
Multiply by T : $=(1)$		

If calculated χ^2 exceeds the tabulated values on line 23 there is evidence for a significant difference.

Probability levels :	90%	95%	99%	99.9%
Tabulated χ^2 :	2.71	3.84	6.64	10.83

Continue only if significance level is sufficiently high.

Confidence limits for the difference between proportions :

If either of the proportions is <0.1 or >0.9 the confidence limits are seriously asymmetrical and the following calculation will be inaccurate.

Calculation of variances within each group

$ca/A =$	$=P_A$	$cb/B =$	$=P_B$	
$1-P_A =$	$=Q_A$	$1-P_B =$	$=Q_B$	
$P_A Q_A/A =$	$=V_A$	$P_B Q_B/B =$	$=V_B$	

Sum of the two variances : $V_A + V_B =$ .... $=V_D$

Standard error of the difference, $\sqrt{V_D} =$ .... $=SE_D$

Difference between the proportions, $P_A - P_B =$ .... $= D$

(because of line 2 above, t will be 2 approximately)

95% confidence limits for the difference between proportions
Upper : $D + 2 \times SE_D =$ ....
Lower : $D - 2 \times SE_D =$

1	COMPARISON OF TWO PROPORTIONS
2	_based on "large" samples, preferably with >30 observations_
3	Nature of the observations: Survival of Aphids. *
4	Group A: Treated with 0.65 Na Oleate.
5	Group B: Treated with 1.10 Na Oleate.
6	Test of difference between the two groups:

		Group A	Group B	Total	
7					
8	No. with the attribute :	13 $=ca$	3 $=cb$	16 $=C$	
9	No. without attribute :	55 $=da$	62 $=db$	117 $=D$	
10	Total no. of observations :	68 $=A$	65 $=B$	133 $=T$	

11	Calculate ⎰	$\dfrac{T\,(|(ca \times db) - (cb \times da)| - T/2)^2}{A \times B \times C \times D}$ as below
12	X^2 from ⎱	

13	$ca \times db =$	806	$A =$	68
14	$cb \times da =$	165	Multiply by B :	4420
15	Subtract and make positive	641	Multiply by C :	70720
16	$0.5 \times T$	66.5	Multiply by D :	8274240 $=(2)$
17	Subtract :	574.5		
18	Square :	330050	$X^2 = \dfrac{(1)}{(2)} =$	
19	Multiply by T :	43896683 $=(1)$		5.31

20	If calculated X^2 exceeds the tabulated values on line 23 there is
21	evidence for a significant difference.

22	Probability levels :	90%	95%	99%	99.9%
23	Tabulated X^2 :	2.71	3.84	6.64	10.83

24	Continue only if significance level is sufficiently high.
25	Confidence limits for the difference between proportions:
26	If either of the proportions is <0.1 or >0.9 the confidence
27	limits are seriously asymmetrical and the following calculation
28	will be inaccurate.

29	Calculation of ⎧	$ca/A =$ 0.191 $=PA$	$cb/B =$ 0.046 $=PB$
30	variances within ⎨	$1-PA =$ 0.809 $=QA$	$1-PB =$ 0.954 $=QB$
31	each group ⎩	$P_A Q_A/A =$ 0.00227 $=VA$	$P_B Q_B/B =$ 0.00068 $=VB$

32	Sum of the two variances : $VA + VB =$ 0.00295 $=VD$
33	Standard error of the difference, $\sqrt{VD} =$ 0.0543 $=SE_D$
34	Difference between the proportions, $PA - PB =$ 0.145 $= D$
35	(because of line 2 above, t will be 2 approximately)
36	95% confidence limits for the ⎰ Upper : $D + 2 \times SE_D =$ 0.254
37	difference between proportions ⎱ Lower : $D - 2 \times SE_D =$ −0.036

* Snedecor & Cochran, Statistical Methods, Iowa State
University Press 1967, p. 218 .

CHI-SQUARED TEST OF CONFORMITY WITH A HYPOTHESIS
("Goodness-of-Fit")

Nature of the observations :

The Hypothesis :

(1) Enter the hypothesis in numerical form in column HYP.
(2) Enter observed occurrences in O, and those expected by
    the hypothesis in E. If any expected value is less than
    5, amalgamate with adjacent groups.
(3) Calculate χ^2 in last three columns. If only two groups exist,
    reduce all $|O-E|$ values by 0.5 before squaring.

Group No.	Description	HYP	O	E	$|O-E| = D$	D^2	$\dfrac{D^2}{E}$
1							
2							
3							
4							
5							
6							
7							
8							
9							
10							
11							
12							
13							
TOTALS				=		$\Sigma\left(\dfrac{D^2}{E}\right) = \chi^2 =$	

No. of groups (lines) in the test, $n = \ldots\ldots$

Number of independent figures ("parameters")
including the total, which had to be taken from $\Big\}$ $d = \ldots\ldots$
the data for constructing the expected values:

Degrees of freedom for the test is $n-d$,   d.f. $= \ldots\ldots$

Tabulated value for χ^2 at this d.f. is : _____

If calculated χ^2 exceeds tabulated for above d.f. then there is
evidence for departure from the hypothesis

STATFORM 17

1	CHI-SQUARED TEST OF CONFORMITY WITH A HYPOTHESIS
2	("Goodness-of-fit")
3	Nature of the observations : Mendel's observations ;
4	colour of peas from dihybrid cross. *
5	The Hypothesis : Independent segregation in ratio
6	9-3-3-1
7	(1) Enter the hypothesis in numerical form in column HYP.
8	(2) Enter observed occurrences in O, and those expected by
9	the hypothesis in E. If any expected value is less than
10	5, amalgamate with adjacent groups.
11	(3) Calculate x^2 in last three columns. If only two groups exist,
12	reduce all $|O-E|$ values by 0·5 before squaring.

	Group No.	Description	HYP	O	E	$\|O-E\| = D$	D^2	$\dfrac{D^2}{E}$
15	1	Round/yellow	9	315	312·75	2·25	5·0625	0·016
16	2	Round/green	3	108	104·25	3·75	14·0625	0·135
17	3	Wrinkled/yellow	3	101	104·25	−3·25	10·5625	0·101
18	4	Wrinkled/green	1	32	34·75	−2·75	7·5625	0·218
19	5							
20	6							
21	7							
22	8							
23	9							
24	10							
25	11							
26	12							
27	13							
28	**TOTALS**		16	556 = 556		$\Sigma\left(\dfrac{D^2}{E}\right) = x^2 = 0.470$		

30	No. of groups (lines) in the test, $n = $ ....4....
31	Number of independent figures ("parameters")
32	including the total, which had to be taken from $d = $ ....1......
33	the data for constructing the expected values:
34	Degrees of freedom for the test is $n-d$, d.f. = ...3......
35	Tabulated value for x^2 at this d.f. is :
36	If calculated x^2 exceeds tabulated for above d.f. then there is
37	evidence for departure from the hypothesis

* R.E. Parker, Introductory Statistics for Biology, Edward Arnold 1973, p.37-38.

STATFORM 18

1	CHI-SQUARED TEST OF CONFORMITY WITH POISSON HYPOTHESIS
2	with provision for two sets of observations.

3 | Observations in f must be counts, not proportions, fractions or percentages.

4	Value	Nature of the observations :					
5	of	A :			B :		
6	count						
7	y	f	$f \times y$	$f \times y^2$	f	$f \times y$	$f \times y^2$
8	0						
9	1						
10	2						
11	3						
12	4						
13	5						
14	6						
15	7						
16	8						
17	9						
18	10						
19	Σ						
20		N	Σy	$\Sigma (y^2)$	N	Σy	$\Sigma (y^2)$

			A	B
21	Calculation of x^2 for data of :			
22	Mean, $\Sigma y / N$,	$\bar{y}$ =		
23	Sum of squared observations, $\Sigma (y^2)$ =			
24	Correction factor $(\Sigma y)^2 / N$,	CF =		
25	Sum-of-squares,	SSy =		
26	Degrees of freedom,	$N-1$ =		
27	Variance, $SS/(N-1)$,	Vy =		
28	x^2 for variance test, $SS/\bar{y}$,	x^2 =		
29	Tabulated values of x^2 at $\{$ P.025 :			
30	the two probability levels $($ P.975 :			

31 | If calculated x^2 lies outside these limits, the observations depart from the Poisson

32	Standard deviation, $\sqrt{\bar{y}}$	SDy =		
33	Standard error of mean, $SD/\sqrt{N}$,	$SE\bar{y}$ =		
34	Student's t at P.	t =		
35	Confidence value, $\pm t \times SE\bar{y}$,	C =		
36	Confidence limits $\{$ Upper : $\bar{y} + c$	=		
37	for the mean $($ Lower : $\bar{y} - c$	=		

STATFORM 18

1	CHI-SQUARED TEST OF CONFORMITY WITH POISSON HYPOTHESIS
2	with provision for two sets of observations.
3	_Observations in f must be counts, not proportions, fractions or percentages._

Line	Value of count	Nature of the observations:					
4		A: Number of weed seeds			B:		
5		in 98 batches. *					
6							
7	y	f	$f \times y$	$f \times y^2$	f	$f \times y$	$f \times y^2$
8	0	3	0	0			
9	1	17	17	17			
10	2	26	52	104			
11	3	16	48	144			
12	4	18	72	288			
13	5	9	45	225			
14	6	3	18	108			
15	7	5	35	245			
16	8	0	0	0			
17	9	1	9	81			
18	10						
19	Σ	98	296	1212			
20		N	Σy	$\Sigma(y^2)$	N	Σy	$\Sigma(y^2)$

Line	Calculation of x^2 for data of :		A	B
21	Calculation of x^2 for data of :		A	B
22	Mean, $\Sigma y / N$.	$\bar{y}$ =	3.02	
23	Sum of squared observations, $\Sigma(y^2)$ =		1212	
24	Correction factor $(\Sigma y)^2/N$, CF =		894.04	
25	Sum-of-squares, SSy =		317.96	
26	Degrees of freedom, $N-1$ =		97	
27	Variance, $SS/(N-1)$, Vy =		3.28	
28	x^2 for variance test, $SS/\bar{y}$, x^2 =		105.28	
29	Tabulated values of x^2 at { P.025 :		>130	
30	the two probability levels (P.975 :		≈ 70	
31	_If calculated x^2 lies outside these limits, the observations depart from the Poisson_			
32	Standard deviation, $\sqrt{\bar{y}}$ SDy =		1.74	
33	Standard error of mean, $SD/\sqrt{N}$, SE$\bar{y}$ =		0.176	
34	Student's t at P.05 t =		1.99	
35	Confidence value, $\pm t \times SE\bar{y}$, C =		0.35	
36	Confidence limits { Upper : $\bar{y}+c$ =		3.37	
37	for the mean (Lower : $\bar{y}-c$ =		2.67	

* Snedecor & Cochran, Statistical Methods, Iowa State University Press 1967, p. 237.

CHI-SQUARED TEST OF 2 × 2 CONTINGENCY TABLE

Nature of the observations :
Column categories A and B :
Row categories   C and D :

Setting up the contingency table.
(1) Enter observed occurrences in O boxes and fill in totals.
(2) Calculate expected occurrences in E boxes. Each E is the product
of its row and column totals, divided by the grand total.
(If any expected value is less than 5 it may be preferable to use
"Fisher's exact test" on Statform 20).
Calculate and enter any one $|O-E|$ difference, then as a
check see that the other three are identical.

Contingency table

Categories ⟶		A		B		Totals
		O	E	O	E	
C	O and E					
	$|O-E|$					
		O	E	O	E	
D	O and E					
	$|O-E|$					
Totals		+		==		

Calculation of χ^2 from : $\Sigma\left[\dfrac{(|O-E|-0.5)^2}{E}\right]$

Box	Expected Values E	1/E	
CA			$|O-E|$ from above : ....
CB			Subtract 0.5 : ....
DA			Squared : ....
DB			Multiply by $\}$ $\chi^2 =$
$\Sigma(1/E)$	=		$\Sigma(1/E)$ to give $\}$ ....

Compare with $\}$ χ^2 table $\}$	Probabilities :	0.10	0.05	0.01	0.001
	χ^2 :	2.71	3.84	6.64	10.83

STATFORM 19

1	CHI-SQUARED TEST OF 2×2 CONTINGENCY TABLE
2	
3	Nature of the observations : **Laterality in 400 children.** *
4	Column categories A and B: **Left and right handed.**
5	Row categories C and D: **Left and right eye dominant**
6	Setting up the contingency table.
7	(1) Enter observed occurrences in O boxes and fill in totals.
8	(2) Calculate expected occurrences in E boxes. Each E is the product
9	of its row and column totals, divided by the grand total.
10	(If any expected value is less than 5 it may be preferable to use
11	"Fisher's exact test" on Statform 20).
12	Calculate and enter any one $|O-E|$ difference, then as a
13	check see that the other three are identical.

Contingency table

Categories ⟶		A		B		Totals
		O	E	O	E	
C	O and E	27	18·5	110	118·5	137
	$|O-E|$	8·5		8·5		
		O	E	O	E	
D	O and E	27	35·5	236	227·5	263
	$|O-E|$	8·5		8·5		
Totals		54	+	346	=	400

Calculation of χ^2 from : $\sum\left[\dfrac{(|O-E| - 0.5)^2}{E}\right]$

Box	Expected Values E	1/E	
CA	18·5	0·05405	$|O-E|$ from above : ......8·5......
CB	118·5	0·00844	Subtract 0·5 : ......8·0......
DA	35·5	0·02817	Squared : ......64·0......
DB	227·5	0·0044	Multiply by } $\chi^2 =$ ......6·084......
$\sum(1/E)$	=	0·09506	$\sum(1/E)$ to give }

Compare with } Probabilities :

χ^2 table }

	0·10	0·05	0·01	0·001
χ^2 :	2·71	3·84	6·64	10·83

* R.E. Parker, Introductory Statistics for Biology, Edward Arnold 1973, p. 41 – 42.

1	FISHER'S EXACT TEST FOR 2 × 2 CONTINGENCY TABLE
2	For cases where there are too few observations for the
3	reliable use of the χ^2 test
4	Nature of the observations:
5	Column categories A and B :
6	Row categories C and D :
7	Contingency table : Enter observed occurrences in boxes
8	ca, cb, da and db and fill in totals.

	A	B	Totals
C	ca	cb	c
D	da	db	d
Totals	a	b	T

18	Calculation of probability of this particular outcome among all
19	possible outcomes with the same marginal totals:

$$P = \frac{a! \times b! \times c! \times d!}{ca! \times cb! \times da! \times db! \times T!}$$

22 If T small, less than 12, use direct calculation :

$$P = \frac{\dots\dots x \dots\dots x \dots\dots x \dots}{\dots\dots x \dots\dots x \dots\dots x \dots\dots x \dots} = \dots\dots$$

25 But if T is larger preferably use logarithms:
26 (eg: Fisher's and Yates' Tables, Table $\underline{XXX}$)

27	Log. ca! = ...	Log. a! = ...
28	Log. cb! = ...	Log. b! = ...
29	Log. da! = ...	Log. c! = ...
30	Log. db! = ...	Log. d! = _____
31	Log. T! = _____	Σ (Log. dividend) = ...
32	Σ (Log. divisor) =	⟶ Subtract ⟶ _____

33 Antilog. LP = ... = P | Log. Probability; LP =
34 P is only the probability of this particular outcome in boxes ca, cb, da and db.
35 If none contain zero, more extreme outcomes are possible. For a
36 complete 1-tailed test of the null hypothesis, their probabilities should also
37 be calculated and added to that in line 33 above.

1	FISHER'S EXACT TEST FOR 2×2 CONTINGENCY TABLE
2	For cases where there are too few observations for the
3	reliable use of the χ^2 test

4 Nature of the observations: **Reaction of 5 cats to medication.**
5 Column categories A and B : **A = recovered , B = remain infected.**
6 Row categories   C and D : **Treated with Dawa, and Controls**

7 Contingency table : Enter observed occurrences in boxes
8 ca, cb, da and db and <u>fill in totals.</u>

		A	B	Totals
Controls →	C	ca 2	cb 3	c 5
Treated with Dawa →	D	da 4	db 1	d 5
	Totals	a 6	b 4	T 10

18 Calculation of probability of this particular outcome among all
19 possible outcomes with the same marginal totals:

$$ P = \frac{a! \times b! \times c! \times d!}{ca! \times cb! \times da! \times db! \times T!} $$

22 If T small, less than 12, use direct calculation :

$$ P = \frac{6! \times 4! \times 5! \times 5!}{2! \times 3! \times 4! \times 1! \times 10!} = 0.238 $$

25 But if T is larger preferably use logarithms:
26 (eg: Fisher's and Yates' Tables, Table <u>XXX</u>)

27	Log. ca!	= 0.301	Log. a!	=	2.857
28	Log. cb!	= 0.778	Log. b!	=	1.38
29	Log. da!	= 1.380	Log. c!	=	2.079
30	Log. db!	= 0.0	Log. d!	=	2.079
31	Log. T!	= 6.56	Σ (Log. dividend)	=	8.395
32	Σ (Log. divisor)	= 9.019	→ Subtract →		9.019

33 Antilog. LP = 0.238 = P | Log. Probability; LP = $\overline{1}$·376

34 P is only the probability of this particular outcome in boxes ca, cb, da and db.
35 If none contain zero, more extreme outcomes are possible. For a
36 complete 1-tailed test of the null hypothesis, their probabilities should also
37 be calculated and added to that in line 33 above.

1	CHI-SQUARED TEST OF $n \times m$ CONTINGENCY TABLE
2	When n or m or both are greater than 2.
3	Nature of the observations:
4	Row categories : Column categories :
5	W: Y: A : C :
6	X: Z: B: D :
7	No. of rows, m = ... No. of columns, n = ...
8	Degrees of freedom for x^2, $(n-1) \times (m-1)$ = ... = d.f.
9	(1) Enter observed occurrences on lines labelled O.
10	(2) Calculate expected values on lines E. Each E is the product of
11	its row and column totals divided by the grand total

Categories ⟶		A	B	C	D	Totals OBS	x^2
	O :						
	E :						
W	O–E :						
	$(O-E)^2$:						
	$(O-E)^2/E$:						
	O :						
	E :						
X	O–E :						
	$(O-E)^2$:						
	$(O-E)^2/E$:						
	O :						
	E :						
Y	O–E :						
	$(O-E)^2$:						
	$(O-E)^2/E$:						
	O :						
	E :						
Z	O–E :						
	$(O-E)^2$:						
	$(O-E)^2/E$:						
Σ { observations :							
$(O-E)^2/E$:							

36	If calculated x^2 exceeds tabulated value at the d.f. entered
37	above, then the categories are not independent x^2

1	CHI-SQUARED TEST OF $n \times m$ CONTINGENCY TABLE							
2	When n or m or both are greater than 2.							
3	Nature of the observations: Ecology of moss. *							
4	Row categories: Humus depth.			Column categories: Moss vigour.				
5	W: <1 cm Y: >2 cm			A: absent C: dense pure weft.				
6	X: 1-2 cm Z:			B: mixed weft D:				
7	No. of rows, m = ...3...			No. of columns, n = ...3...				
8	Degrees of freedom for x^2, $(n-1) \times (m-1)$ = ...4... = d.f.							
9	(1) Enter observed occurrences on lines labelled O.							
10	(2) Calculate expected values on lines E. Each E is the product of							
11	its row and column totals divided by the grand total							

12–13 Categories ⟶		A	B	C	D	Totals OBS	x^2
14	O:	36	15	7			
15	E:	27.84	17.79	12.37		58	
16 W	O−E:	8.16	−2.79	−5.37			
17	$(O-E)^2$:	66.59	7.78	28.84			
18	$(O-E)^2/E$:	2.39	0.44	2.33			5.16
19	O:	65	37	44			
20	E:	70.08	44.77	31.15		146	
21 X	O−E:	−5.08	−7.77	12.85			
22	$(O-E)^2$:	25.81	60.37	165.12			
23	$(O-E)^2/E$:	0.37	1.35	5.30			7.02
24	O:	43	40	13			
25	E:	46.08	29.44	20.48		96	
26 Y	O−E:	−3.08	10.56	−7.48			
27	$(O-E)^2$:	9.49	111.51	55.95			
28	$(O-E)^2/E$:	0.21	3.79	2.73			6.73
29	O:						
30	E:						
31 Z	O−E:						
32	$(O-E)^2$:						
33	$(O-E)^2/E$:						
34 Σ { observations:		144	92	64		300	
35 { $(O-E)^2/E$:		2.97	5.58	10.36			18.91

36	If calculated x^2 exceeds tabulated value at the d.f. entered }
37	above, then the categories are not independent } x^2

* R.E. Parker, Introductory Statistics for Biology, Edward Arnold 1973, p. 53.

1	MANN – WHITNEY U-TEST BETWEEN TWO SMALL SAMPLES
2	For samples exceeding 20 observations an approximate
3	normal deviate test or a t-test may be appropriate.
4	Nature of the observations :
5	A :
6	B :
7	Ordering and ranking the observations :
8	(1) List observations in order of magnitude within each group,
9	lesser values first, on lines labelled OBS.
10	(2) Assign ascending ranks 1, 2, 3, ... to the whole set of observations
11	on lines labelled RNK. Ranks must continue to rise even
12	through a run of identical observations (known as "ties").
13	(3) If ties occur between the groups, amend the ranks in any such
14	run of ties to the mean rank within that run.
15	(4) Enter the number of observations (n) and add the ranks (R),
16	adjusted for ties if necessary, for each group.

17	A							$n_A =$
18	OBS:							
19	RNK:							$R_A =$
20								
21	B							$n_B =$
22	OBS:							
23	RNK:							$R_B =$
24								

25	Check $\left\{\begin{array}{l} R_A + R_B = 0.5 \times (n_A + n_B) \times (n_A + n_B + 1) \, ? \\ = 0.5 \times (\quad) \times (\quad) \, ? \end{array}\right.$
26	
27	Lesser R must be ≤ tabulated rank-sum for significant difference

28	Calculation of U :	U_A		U_B	
29	Enter $n_A \times n_B$:				
30	Enter $n_x + 1$:	$n_A + 1 =$		$n_B + 1 =$	
31	Multiply by n_x :	$\times n_A =$		$\times n_B =$	
32	Divide by 2 :	$\div 2 =$		$\div 2 =$	
33	Add $n_A \times n_B$:	$+ (n_A \times n_B) =$		$+ (n_A \times n_B) =$	
34	Subtract R_x ⎰ :	$- R_A =$	$= U_A$	$- R_B =$	$= U_B$
35	to give U_x ⎱				
36	Check : $U_A + U_B = n_A \times n_B$?				
37	Lesser U must be ≤ Wilcoxon's tabulated U for significant difference.				

1	MANN–WHITNEY U-TEST BETWEEN TWO SMALL SAMPLES
2	For samples exceeding 20 observations an approximate
3	normal deviate test or a t-test may be appropriate.
4	Nature of the observations: Textbook example, modified.*
5	A : x - values
6	B : y - values (with highest value omitted)
7	Ordering and ranking the observations :
8	(1) List observations in order of magnitude within each group,
9	lesser values first, on lines labelled OBS.
10	(2) Assign ascending ranks 1, 2, 3, ... to the whole set of observations
11	on lines labelled RNK. Ranks must continue to rise even
12	through a run of identical observations (known as "ties").
13	(3) If ties occur between the groups, amend the ranks in any such
14	run of ties to the mean rank within that run.
15	(4) Enter the number of observations (n) and add the ranks (R),
16	adjusted for ties if necessary, for each group.

	A									
18	OBS:	4	5	6	8	11	14		$n_A = 6$
19	RNK:								
20		1	2	3	4.5	9 8	11		$R_A = 29$

	B							
22	OBS:	8	9	11	11	13		$n_B = 5$
23	RNK:	5	6	7	8	10		
24				8	8			$R_B = 37$

25	Check $\{$ $R_A + R_B = 0.5 \times (n_A + n_B) \times (n_A + n_B + 1)$?
26	$66 = 0.5 \times (11) \times (12)$?
27	Lesser R must be $\leq$ tabulated rank-sum for significant difference

	Calculation of U :	U_A		U_B	
29	Enter $n_A \times n_B$:		30		30
30	Enter $n_x + 1$:	$n_A + 1$ =	7	$n_B + 1$ =	6
31	Multiply by n_x :	$\times n_A$ =	42	$\times n_B$ =	30
32	Divide by 2 :	$\div 2$ =	21	$\div 2$ =	15
33	Add $n_A \times n_B$:	$+(n_A \times n_B)$ =	51	$+(n_A \times n_B)$ =	45
34	Subtract R_x $\}$:	$- R_A =$		$- R_B =$	
35	to give U_x		22 $= U_A$		8 $= U_B$

36	Check : $U_A + U_B = n_A \times n_B$? $22 + 8 = 5 \times 6$
37	Lesser U must be $\leq$ Wilcoxon's tabulated U for significant difference.

* R. C. Campbell, Statistics for Biologists, C.U.P. 1974, p. 59.

1	WILCOXON SIGNED-RANK TEST FOR PAIRED COMPARISON
2	For samples exceeding 25 pairs an approximate
3	normal deviate, x^2 or t-test may be appropriate.
4	Nature of the observations :
5	A :
6	B :
7	Number of pairs, $n =$
8	Ranking the differences within pairs :
9	(1) List the observations and differences within pairs in the
10	first five columns. Omit zero differences.
11	(2) Assign ascending ranks 1, 2, 3,..... to the whole set of differences.
12	Ranks must continue to rise even through a run of identical
13	differences, known as "ties.
14	(3) If ties occur between positive and negative columns, amend ranks
15	in any such run of ties to the mean rank within the run.

Pair No.	Observations A	B	Differences A–B Positive	Negative	Ranks of differences Positive	Negative
1						
2						
3						
4						
5						
6						
7						
8						
9						
10						
11						
12						
13						
14						
15						

Sums of ranks : $R_P =$ $R_N =$

Check $\begin{cases} R_P + R_N & = 0.5 \times n(n+1) & ? \\ & = \end{cases}$

Compare lesser R with Wilcoxon's tabulated signed rank statistic T. Lesser R must be $\leq$ T for significant difference.

STATFORM 23

1	WILCOXON SIGNED-RANK TEST FOR PAIRED COMPARISON
2	For samples exceeding 25 pairs an approximate
3	normal deviate, x^2 or t-test may be appropriate.
4	Nature of the observations: Colour fastness of dyes.*
5	A : dye A
6	B : dye B
7	Number of pairs, n = 9 (identical pairs omitted, i.e. pair no. 6)
8	Ranking the differences within pairs :
9	(1) List the observations and differences within pairs in the
10	first five columns. Omit zero differences.
11	(2) Assign ascending ranks 1, 2, 3,..... to the whole set of differences.
12	Ranks must continue to rise even through a run of identical
13	differences, known as "ties.
14	(3) If ties occur between positive and negative columns, amend ranks
15	in any such run of ties to the mean rank within the run.

Pair No.	Observations		Differences A–B		Ranks of differences	
	A	B	Positive	Negative	Positive	Negative
1	4	7		3		5
2	4	6		2		~~4~~ 3.5
3	5	9		4		7
4	1	5		4		8
5	3	6		3		6
6	3	3	—	—	—	—
7	5	4	1		~~1~~ 1.5	
8	6	7		1		~~2~~ 1.5
9	6	4	2		~~4~~ 3.5	
10	3	10		4		9
11						
12						
13						
14						
15						
			Sums of ranks :		$R_P = 5$	$R_N = 40$

34	Check $\begin{cases} R_P + R_N = 0.5 \times n(n+1) \\ 5 + 40 = 0.5 \times 9(10) = 45? \end{cases}$
36	Compare lesser R with Wilcoxon's tabulated signed rank statistic T.
37	Lesser R must be $\leq$ T for significant difference.

* R.C. Campbell, Statistics for Biologists, C.U.P 1974, p.65

STATFORM 24

KRUSKAL — WALLIS TEST
For differences between medians of several samples.
(Non parametric analogue of Statform 7)

Nature of the observations:

Ranking the observations:
(1) List observations in treatment columns, lowest values first.
(2) Assign ranks 1, 2, 3,..... to the whole set. Ranks must continue to rise even through a run of identical observations (ties).
(3) Amend ranks within each tied run to the mean rank in the run.

Treatment	OBS	Ranks	OBS	Ranks	OBS	Ranks	OBS	Ranks	Totals
									N=
n =									
$R(sum)$=									
R^2 =									
R^2/n =									

Check $\left\{ \begin{array}{l} \Sigma R = 0.5 \times (N^2 + N) \\ \quad = \quad \times (\quad) \end{array} \right.$?

Calculation of H:
Enter $\Sigma (R^2/n)$: ....
Divide by N (=) : ....
Divide by $N+1$ (=) : ....
Multiply by 12 : ....
Subtract $3(N+1)$ (=) $\Big\}$: H ....
to give statistic H

(Refer H to table of the H Statistic

With more than 3 treatments or more than 5 observations in a treatment, refer H to a χ^2 table with degrees of freedom one less than the number of treatments.

STATFORM 24

1	KRUSKAL – WALLIS TEST
2	For differences between medians of several samples.
3	(Non parametric analogue of Statform 7)

4 Nature of the observations :

5 **Textbook example of rank observations.** *

6 Ranking the observations :

7 (1) List observations in treatment columns, lowest values first.

8 (2) Assign ranks 1, 2, 3, ... to the whole set. Ranks must continue to

9 rise even through a run of identical observations (ties).

10 (3) Amend ranks within each tied run to the mean rank in the run.

Treatment	A		B		C				Totals
	OBS	Ranks	OBS	Ranks	OBS	Ranks	OBS	Ranks	
13	8	7	3	1.5	3	1.5			
14	8	7	4	3	6	5			
15	12	11	5	4	8	7			
16	14	12	9	9	10	10			
17	15	13							
$n =$	5		4		4			$N =$	13
$R (sum) =$		50		17.5		23.5			91
$R^2 =$		2500		306.25		552.25			
$R^2/n =$		500		76.56		138.06			714.62

26 Check $\left\{ \begin{array}{l} \sum R = 0.5 \times (N^2 + N) \\ 91 = 0.5 \times (169 + 13) \end{array} \right.$?

28 Calculation of H :

29   Enter $\sum (R^2/n)$:   714.62

30   Divide by N (= 13) :   54.97

31   Divide by N+1 (= 14) :   3.93

32   Multiply by 12 :   47.16   (Refer H to

33   Subtract 3(N+1) (= 42))   : $H = 5.16$ table of the

34   to give statistic H   H Statistic)

35 With more than 3 treatments or more than 5 observations in a

36 treatment, refer H to a χ^2 table with degrees of freedom

37 one less than the number of treatments.

* R.C. Campbell, Statistics for Biologists, C.U.P. 1974 p. 63.

FRIEDMANN'S TEST FOR RANDOMIZED BLOCK
(Non parametric analogue of Statform 10.)

Nature of the observations :

Ranking the observations :

(1) Tabulate the observations in OBS columns by treatment and block.

(2) Assign ranks $1, 2, 3, \ldots$ to the observations in column R separately within each block. Ranks must continue to rise even through a run of identical observations ("ties")

(3) Amend ranks within each tied run to the mean rank of the run.

No. of treatments, $K =$ ___   No. of blocks, $n =$ ___

Trtmt:														
Block	OBS	R	OBS	R	OBS	R	OBS	R	OBS	R	OBS	R
1													
2													
3													
4													
5													
6													

$\Sigma R =$

$(\Sigma R)^2 =$

Check $\left\{ \begin{array}{l} \Sigma R = 0.5 \times n \times K \times (K+1) \\ = \quad \times \quad \times \quad \times (\quad) \end{array} \right.$?

Calculation of S :

$\Sigma (R^2) =$ ....

$(\Sigma R)^2 / K =$ ___

Subtract to give Friedmann's $S =$ ___

Refer to table of Friedmann's S - statistic, but if either K or n values exceed those of the table, estimate X^2 as $\dfrac{6 \times S}{\Sigma R} = \dfrac{6 \times \_\_}{\_\_} =$ ....

and refer to tabulated X^2 with $K-1$ degrees of freedom.

STATFORM 25

1	FRIEDMANN'S TEST FOR RANDOMIZED BLOCK
2	(Non parametric analogue of Statform 10.)
3	Nature of the observations : Textbook example ; five scores
4	in each of four treatments. *
5	Ranking the observations :
6	(1) Tabulate the observations in OBS columns by treatment and block.
7	(2) Assign ranks 1, 2, 3, ... to the observations in column R separately
8	within each block. Ranks must continue to rise even through a
9	run of identical observations ("ties")
10	(3) Amend ranks within each tied run to the mean rank of the run.
11	No. of treatments. $K = 4$ No. of blocks, $n = 5$

Trtmt:	A		B		C		D					
Block	OBS	R	OBS	R	OBS	R	OBS	R	OBS	R	OBS	R
1	6	1	8	2	11	3	12	4				
2	2	1	5	2	10	3.5	10	3.5				
3	5	1.5	5	1.5	9	3	11	4				
4	1	1	5	2	8	3	9	4				
5	3	1	7	2.5	8	4	7	2.5				
6												
$\Sigma R =$	5.5		10		16.5		18					
$(\Sigma R)^2 =$	30.25		100		272.25		324					

28	Check $\left\{ \begin{array}{l} \Sigma R = 0.5 \times n \times K \times (K+1) \\ 50 = 0.5 \times 5 \times 4 \times (5) \end{array} \right.$?
30	Calculation of S : $\Sigma (R^2) = $ 726.5
31	$(\Sigma R)^2 / K = $ 625
32	Subtract to give Friedmann's $S = $ 101.5
33	Refer to table of Friedmann's S – statistic, but if
34	either K or n values exceed those of the table,
35	estimate χ^2 as $\dfrac{6 \times S}{\Sigma R} = \dfrac{6 \times 101.5}{50} = $
36	12.18
37	and refer to tabulated χ^2 with $K-1$ degrees of freedom.

* R. C. Campbell, Statistics for Biologists, C.U.P. 1974, p. 70 .

1	SPEARMAN RANK CORRELATION (allowing for tied ranks)
2	Nature of the observations:
3	A : B :
4	(1) List the observations in pairs in columns 2 and 3.
5	(2) Assign ranks 1,2,3,..... separately to A and B in cols. 4 and 5. Ranks
6	must continue rising through any runs of identical observations ("ties").
7	(3) If ties exist, amend ranks in tied runs to the mean rank within the run.
8	(4) Enter differences between ranks in column 6, and their squares in column 7.
9	(5) List the number of ties in each run of ties in columns 8 and 9.

Line	Pair No.	Observations 2. A	3. B	Ranks 4. A	5. B	Difference 6. A-B	D^2 7.	Ties in each run. 8 T(A)	9 T(B)
10, 11									
12	1								
13	2								
14	3								
15	4								
16	5								
17	6								
18	7								
19	8								
20	9								
21	10								
22	11								
23	12								
24	13								

25	No. of pairs of observations, n = ΣD^2 =
26	$N' = n^3 - n$ = For cols. 8 and 9 $\Big\}$ T_A =
27	.... Calculate $\Sigma(T^3 - T)$ $\Big\}$ T_B =
28	$SS_A = (N' - T_A)/12$ = $SS_B = (N' - T_B)/12$ =
29	If there are no ties :
30	Spearman rank $\Big\{$ $r_S = 1 - \dfrac{6 \times \Sigma D^2}{N'}$ = $1 - \underline{\qquad}$ =
31	correlation
32	If ties exist :
33	$r_S = \dfrac{SS_A + SS_B - \Sigma D^2}{2 \times \sqrt{SS_A \times SS_B}}$ = $\underline{\qquad}$ =
34	
35	Refer r_S to table of Spearman coefficient, or if n too large
36	calculate : $t = r_S \times \sqrt{\dfrac{n-2}{1-r_S^2}}$ = $\Big\}$ refer to t-table
37	$\Big\}$ for $n-2$ d.f.

STATFORM 26

1	SPEARMAN RANK CORRELATION (allowing for tied ranks)

2 Nature of the observations: weaning weight on fattening period.*

3 A: weight in pounds.   B: period in days.

4 (1) List the observations in pairs in columns 2 and 3.

5 (2) Assign ranks 1, 2, 3, ... separately to A and B in cols. 4 and 5. Ranks

6 must continue rising through any runs of identical observations ("ties").

7 (3) If ties exist, amend ranks in tied runs to the mean rank within the run.

8 (4) Enter differences between ranks in column 6, and their squares in column 7.

9 (5) List the number of ties in each run of ties in columns 8 and 9.

10 Pair	Observations		Ranks		Difference	D^2	Ties in each run.	
11 No.	2. A	3. B	4. A	5. B	6. A-B	7.	8 T(A)	9 T(B)
12 1	39	142	1	9	-8	64		
13 2	41	147	2	10	-8	64		
14 3	49	119	5	7	-2	4		
15 4	50	117	~~7~~ 6.5	6	0.5	0.25	2	
16 5	57	111	9	~~4~~ 3	6	36		3
17 6	56	111	8	~~2~~ 3	5	25		
18 7	59	105	10	1	9	81		
19 8	50	111	~~6~~ 6.5	~~3~~ 3	3.5	12.25		
20 9	46	115	~~3~~ 3.5	5	-1.5	2.25	2	
21 10	46	121	~~4~~ 3.5	8	-4.5	20.25		
22 11								
23 12								
24 13								

25 No. of pairs of observations, n = 10 $\Sigma D^2 = 309$

26 $N' = n^3 - n = $ .....990.....

27 For cols. 8 and 9 Calculate $\Sigma (T^3 - T)$ $T_A = 12$ $T_B = 24$

28 $SS_A = (N' - T_A)/12 = $ $SS_B = (N' - T_B)/12 = $

29 If there are no ties: (calculation inserted for example only)

30 Spearman rank $\left\{ \right.$ correlation

31 $r_S = 1 - \dfrac{6 \times \Sigma D^2}{N'} = 1 - \dfrac{1854}{990} = -0.873$

32 If ties exist:

33 $r_S = \dfrac{SS_A + SS_B - \Sigma D^2}{2 \times \sqrt{SS_A \times SS_B}} = \dfrac{-147}{2 \times 80.998} = -0.907$

35 Refer r_S to table of Spearman coefficient, or if n too large

36 calculate: $t = r_S \times \sqrt{\dfrac{n-2}{1 - r_S^2}} = -6.09$ $\left\{ \right.$ refer to t-table

37 for n-2 d.f.

* R.C. Campbell, Statistics for Biologists, C.U.P 1974, p. 106.

STATFORM 27

1	KENDALL RANK CORRELATION (allowing for tied ranks)
2	Nature of the observations :
3	A : B :
4	Listing and ranking the observations : No. of pairs of observations, $n =$
5	(1) List observations of A increasing downwards in col. 2. Their ranks are in col. 4.
6	(2) Enter corresponding observations of B in col. 3, then assign ranks 1, 2, 3, ····· to
7	them in col. 5, rising even through runs of identical values ("ties").
8	(3) If ties exist, amend all ranks within tied runs to the mean rank in each run.
9	(4) On each line of col. 6, enter no. of B ranks below the line greater than the B
10	rank on the line, but omitting any whose A partner is tied with A on the
11	reference line. Col. 7 similarly, but for B ranks less than that on the line.
12	(5) In cols. 8 and 9, list the number of A and B ties in each run of ties

Pair	Observations 2. A	3. B	Ranks 4. A	5. B	Tally of B ranks 6. Greater	7. Lesser	List of tie runs 8. A ties	9. B ties
			1					
			2					
			3					
			4					
			5					
			6					
			7					
			8					
			9					
			10					
			11					
			12					

Totals :

$S = \Sigma(\text{Greater}) - \Sigma(\text{Lesser}) =$ $T_A :$ $T_B :$

In columns 8 and 9, enter $\Sigma[T(T-1)]$:

$n(n-1) =$	$n(n-1) =$	$N_A \times N_B =$ ....
Less T_A : _____	Less T_B : _____	Square-rooted : ....
gives $N_A =$	gives $N_B =$	Divided by 2, $D =$

Kendall rank correlation $\tau = S/D =$ ..../........ =

Significance : If $n \leqslant 10$, refer to tables of Kendall's S statistic.

If $n > 10$, calculate z and refer to t-table at d.f. ∞; treat as 1-tailed test

$$Z = \tau \left/ \sqrt{\frac{4n+10}{9n(n-1)}} \right. = \ldots\ldots / \sqrt{\frac{\quad}{\quad}} = \ldots\ldots$$

STATFORM 27

1	KENDALL RANK CORRELATION (allowing for tied ranks) *

2 *Nature of the observations* : weaning weight on fattening period.

3 A : weight in pounds.   B : period in days.

4 Listing and ranking the observations : *No. of pairs of observations, n = 10*

5 (1) List observations of A increasing downwards in col. 2. Their ranks are in col. 4.

6 (2) Enter corresponding observations of B in col. 3, then assign ranks 1, 2, 3, ... to

7   them in col. 5, rising even through runs of identical values ("ties").

8 (3) If ties exist, amend all ranks within tied runs to the mean rank in each run.

9 (4) On each line of col. 6, enter no. of B ranks below the line greater than the B

10   rank on the line, but omitting any whose A partner is tied with A on the

11   reference line. Col. 7 similarly, but for B ranks less than that on the line.

12 (5) In cols. 8 and 9, list the number of A and B ties in each run of ties

	Pair	Observations		Ranks		Tally of B ranks		List of tie runs	
		2. A	3. B	4. A	5. B	6. Greater	7. Lesser	8. A ties	9. B ties
15		39	142	1	9	1	8		
16		41	147	2	10	0	8		
17		46	121	3.5	8	0	6	2	
18		46	115	3.5	5	2	4		
19		49	119	5	7	0	5		
20		50	117	6.5	6	0	3	2	
21		50	111	6.5	3	0	1		3
22		56	111	8	3	0	1		
23		57	111	9	3	0	1		
24		59	105	10	1	0	0		
25				11					
26				12					
27					Totals :	3	37		

28 $S = \Sigma (Greater) - \Sigma (Lesser) =$ **−34** T_A: T_B:

29   In columns 8 and 9, enter $\Sigma [T(T-1)]$:   **4**   **6**

30	$n(n-1) =$ 90	$n(n-1) =$ 90	$N_A \times N_B =$ 7224
31	Less T_A : 4	Less T_B : 6	Square-rooted : 84·99
32	gives $N_A =$ 86	gives $N_B =$ 84	Divided by 2, D = 42·5

33 Kendall rank correlation $T = S/D = -34 / 42.05 = -0.8$

34 Significance : If $n \leqslant 10$; refer to tables of Kendall's S statistic.

35 If $n > 10$, calculate z and refer to t-table at d.f. ∞; treat as 1-tailed test

36,37 $Z = T \Big/ \sqrt{\dfrac{4n+10}{9n(n-1)}} = -0.8 \Big/ \sqrt{\dfrac{40+10}{90(9)}} = 3.22$ ** **

* R.C. Campbell, Statistics for Biologists, C.U.P 1974, p. 106.

STATFORM 28

#	
1	BINOMIAL DISTRIBUTION; PREDICTION OF FREQUENCIES
2	When N equal-sized groups each contain n items, of which a
3	mean proportion p possesses one of two alternative attributes.
4	Nature of groups (cells, batches, plots, periods) :
5	
6	Nature of items (objects, events) and attributes :
7	

#		
8	Size of groups, i.e. number of items per group, $n =$	....
9	Mean proportion of items with attribute .... $p =$	....
10	Total number of groups, if known :	
11	But for prediction of $\begin{cases} \text{proportions, take } N = 1 \\ \text{percentages, } N = 100 \end{cases}$ $N =$	....
12		

#	
13	Calculation of binomial frequencies : $q = 1 - p =$
14	(1) Rule across table under value of n in column x.
15	(2) Enter $n-x$ values. (3) Calculate p^x values.
16	(4) Construct Pascal coefficients sufficient for third column.
17	(5) Calculate $q^{(n-x)}$ values and proceed to proportion and
18	number columns as shown by arithmetic signs.

Composition of group		Pascall Coefficient	p^x	$q^{(n-x)}$	Groups with composition x	
x	$n-x$	$\times$ ____	$\times$ ____	$=$ ____	Proportion	Number $\times N =$ ____
0		1	1			
1						
2						
3						
4						
5						
6						
7						
8						
9						
10						
11						
12						
13						

#	
36	Check $\begin{cases} \text{Proportion column sums to } 1.0 : \underline{\hspace{2cm}} \\ \text{Number column sums to } N : \end{cases}$

STATFORM 28

1	BINOMIAL DISTRIBUTION; PREDICTION OF FREQUENCIES
2	When N equal-sized groups each contain n items, of which a
3	mean proportion p possesses one of two alternative attributes.

4	Nature of groups (cells, batches, plots, periods) : 100 Batches of
5	five random digits, which vary from zero to 9. *
6	Nature of items (objects, events) and attributes : Groups in
7	which zero or figure one occur.

8	Size of groups, i.e. number of items per group, $n =$ ....5....
9	Mean proportion of items with attribute ..0 or 1......... $p = $ ..0.2...
10	Total number of groups, if known :
11	But for prediction of $\begin{Bmatrix} \text{proportions, take } N = 1 \\ \text{percentages, } N = 100 \end{Bmatrix}$ $N = $ .....100....
12	

13	Calculation of binomial frequencies :	$q = 1 - p = $ 0.8

14	(1) Rule across table under value of n in column x.
15	(2) Enter n-x values. (3) Calculate p^x values.
16	(4) Construct Pascal coefficients sufficient for third column.
17	(5) Calculate $q^{(n-x)}$ values and proceed to proportion and
18	number columns as shown by arithmetic signs.

Composition of group		Pascall Coefficient	p^x	$q^{(n-x)}$	Groups with composition x	
					Proportion	Number
x	$n-x$	___ X	___ X	___ =	___ $\times N=$	
0	5	1	1	0.3277	0.3277	32.77
1	4	5	0.2	0.4096	0.4096	40.96
2	3	10	0.04	0.512	0.2048	20.48
3	2	10	0.008	0.64	0.0512	5.12
4	1	5	0.0016	0.8	0.0064	0.64
5	0	1	0.00032	1	0.00032	0.03
6						
7						
8						
9						
10						
11						
12						
13						

36	Check $\begin{cases} \text{Proportion column sums to } 1.0 : \\ \text{Number column sums to } N : \end{cases}$	1.00002
37		100

* Snedecor & Cochran, Statistical Methods, Iowa State University Press 1967, p. 206.

	POISSON DISTRIBUTION; PREDICTION OF FREQUENCIES
1	
2	When items occur randomly over a series of N "cells"
3	at a mean density of m items per cell.
4	Nature of cells (plots, periods, groups):
5	
6	Nature of items (objects, events):
7	

8	Total number of cells, if known:	
9	But, for prediction of { proportions, take $N=1$	$N =$ ....
10	percentages, $N=100$	
11	Mean density of items over } $m =$	
12	the whole population of cells }	
13	From table of natural logarithms } $e^{-m} =$	
14	or from $1/(\text{antilog}_{10}(m/2\cdot3))$ }	
15	The calculations which follow are hazardous without a calculator	
16	or logarithms. The Statform assumes a calculator but if	
17	logs are used: $\text{Log}_{10}(e^{-m}) = \dfrac{-m}{2\cdot3}$	
18		

Items per cell	m^x	$m^x \times e^{-m}$ $\div$	$x!$	Cells containing x Proportion $\times N =$	Number
0	1		1		
1			1		
2			2		
3			6		
4			24		
5			120		
6			720		
7			5040		
8			40320		
9			36288×10		
10			36288×10^2		
11			39917×10^3		
12			47900×10^4		
13			62270×10^5		

36	Check { Proportions add to 1·0 approx:
37	Numbers add to N approx:

POISSON DISTRIBUTION; PREDICTION OF FREQUENCIES

When items occur randomly over a series of N "cells"
at a mean density of m items per cell.

Nature of cells (plots, periods, groups):

7 gram samples of Phleum seed; 98 such samples. *

Nature of items (objects, events):

Number of noxious weed-seeds per sample.

Total number of cells, if known:

But, for prediction of $\begin{cases} \text{proportions, take } N = 1 \\ \text{percentages, } N = 100 \end{cases}$ $N = \underline{98}$

Mean density of items over $\Big\}$ $m = \underline{3.02}$
the whole population of cells

From table of natural logarithms $\Big\}$ $e^{-m} = \dfrac{1}{\underline{20.49}}$
or from $1/(\text{antilog}_{10}(m/2.3))$

The calculations which follow are hazardous without a calculator
or logarithms. The Statform assumes a calculator but if
logs are used: $\text{Log}_{10}(e^{-m}) = \dfrac{-m}{2.3} = 0.0488$

Items per cell	m^x	$m^x \times e^{-m}$	$x!$	Cells containing X Proportion	Number
		$\div$		$=$ $\times N =$	
0	1	.0488	1	.0488	4.8
1	3.02	.1474	1	.1474	14.4
2	9.12	.4451	2	.2226	21.8
3	27.54	1.344	6	.224	22.0
4	83.18	4.059	24	.1691	16.6
5	251.21	12.259	120	.1022	10.0
6	758.65	37.022	720	.0514	5.0
7	2291.12	111.8067	5040	.0222	2.2
8	6919.19	337.6565	40320	.0084	.8
9	20895.97	1019.7233	36288 × 10	.0028	.3
10			36288 × 10²		
11			39917 × 10³		
12			47900 × 10⁴		
13			62270 × 10⁵		

Check $\begin{cases} \text{Proportions add to } 1.0 \text{ approx:} & .9989 \\ \text{Numbers add to } N \text{ approx:} & 97.9 \end{cases}$

* Snedecor & Cochran, Statistical Methods, Iowa State University Press 1967, p. 224.

Index